Robotics and Automation Solutions for Inspection and Maintenance in Critical Infrastructures

Konstantinos Loupos
(Editor)

now
the essence of knowledge

Published, sold and distributed by:
now Publishers Inc.
PO Box 1024
Hanover, MA 02339
United States
Tel. +1-781-985-4510
www.nowpublishers.com
sales@nowpublishers.com

Outside North America:
now Publishers Inc.
PO Box 179
2600 AD Delft
The Netherlands
Tel. +31-6-51115274

ISBN: 978-1-63828-282-2
E-ISBN: 978-1-63828-283-9
DOI: 10.1561/9781638282839

Suggested citation: Konstantinos Loupos (ed.). (2024). *Robotics and Automation Solutions for Inspection and Maintenance in Critical Infrastructures*. Boston–Delft: Now Publishers

Table of Contents

Preface xiii

Chapter 1 Introduction to Inspection and Maintenance with Robotics 1

By Konstantinos Loupos

1.1 Introduction 1
1.1.1 Inspection and Maintenance Challenges 3
1.2 Added Value of Robotic and Automation Systems Over Traditional Inspection Approaches 9
1.3 Market Overview and Positioning 11
1.4 Conclusions 12
References 13

Chapter 2 Integrated Inspection and Maintenance Platform for Critical Infrastructures 17

By Georgios Sarigiannis, Athanasios Karydis, Konstantinos Loupos, Eleni Voulgari, Aikaterini Karagiannopoulou, Konstantinos Routsis, Valantis Tsiakos, Georgios Tsimiklis, Angelos Amditis, Marco Antonio Montes Grova, Joaquín Borrego Villa, Francisco Javier Pérez Grau and Antidio Viguria

2.1 Introduction 18
2.2 A Scalable Data Management System Facilitating the Autonomous Inspection and Maintenance of Structural Health Monitoring Operations 19
2.2.1 Introduction 19
2.2.2 Development Framework of the I&M PILOTING Data Management System 22

2.2.2.1 System architectural overview 22
2.2.2.2 A three-phase methodology for the design of the DMS data model 23
2.2.2.3 The standardised OGC-Compliant SensorThings data model 24
2.2.3 The Data Management System Components 26
2.2.3.1 PostgreSQL relational database for data management (RDBMS) 26
2.2.3.2 File management service 27
2.2.3.3 Mission component service 27
2.2.3.4 Mission- & asset-centric data storage/indexing and validation service 28
2.2.3.5 AI metadata service 29
2.2.3.6 Static sensors storage service 29
2.2.4 Authentication, Authorisation and User Management Access 30
2.2.5 Implementation and Evaluation 32
2.2.5.1 Implementation and hosting 32
2.2.5.2 Performance evaluation 33
2.2.6 Software and Data Availability Statement 34
2.3 Data Distribution and Harmonization Layer (DDHL) 34
2.3.1 Introduction 34
2.3.2 Development Framework of the I&M PILOTING Data Distribution and Harmonization Layer 35
2.3.2.1 PILOTING workflow 35
2.3.3 Data Distribution & Harmonization Layer Development 40
2.3.4 Data Distribution & Harmonization Layer Architecture . 42
2.3.4.1 Integration with the internal components of I&M PILOTING platform 42
2.3.4.2 Integration with the external components 44
2.3.4.2.1 Integration with robotic control systems 44
2.3.4.2.2 Integration with AI systems 46
2.3.4.2.3 Integration with IoT wireless static sensors network & SensorThings API 48
2.3.4.3 General framework 48
2.3.4.3.1 Refinery case 48
2.3.4.3.2 Viaduct case 50

2.3.5 Software and Data Availability Statement 51
2.4 The General Robot Control Station 51
2.4.1 Software Modules 51
2.4.1.1 Data model 51
2.4.1.2 Core 52
2.4.1.3 Communications 52
2.4.1.4 View 53
2.4.2 Communications Between Modules 53
2.4.3 Design of Graphical User Interface 54
2.4.4 Communication Interface with Robotic Systems........ 55
2.4.5 Software and Data Availability Statement 56
2.5 Conclusion 56
Acknowledgements 57
References 57

Chapter 3 AI for Corrosion Detection in Oil and Gas Refineries **61**

By Leonidas Valavanis, Georgios Sarigiannis, Athanasios Karydis, Konstantinos Loupos and Ioanna Fergadiotou

3.1 Introduction 62
3.2 Methods 63
3.2.1 CycleGANs 63
3.2.2 Corrosion Detection 64
3.2.2.1 Corrosion detection on pipes 65
3.2.2.1.1 YOLO 65
3.2.2.1.2 MASKRCNN 66
3.2.2.1.3 Swin transformers 67
3.2.2.2 Corrosion detection inside tanks 69
3.3 Dataset Synthesis 69
3.4 Results 72
3.4.1 CycleGANs 73
3.4.2 Corrosion Detection on Pipes 76
3.4.3 Corrosion Detection Inside Tanks 81
3.5 Conclusions 85
Acknowledgements 85
References 85

Chapter 4 Motorway Tunnels Structural Deformation Assessment Using Laser Scanning **87**

By Andreas El Saer, Konstantinos Loupos, George Sarigiannis, Valia Konstantopoulou, Athanasios Karydis and Gerasimos Kouloumbis

4.1 Introduction 88

4.1.1 Ageing Infrastructures ... 88
4.1.2 The PILOTING Project ... 90
4.2 EU Corridors and Roadways Challenges ... 91
4.2.1 Tunnel Inspections – Challenges and Existing Approaches ... 91
4.2.2 Tunnel Inspection – Baseline Requirements ... 92
4.3 Laser Scanning for Tunnel Deformation Detection ... 93
4.3.1 Equipment Capabilities and Setup ... 96
4.3.2 Technical Approach ... 97
4.4 Work Novelty and Industrial Uptake ... 98
4.4.1 Extensions and Digital Assets ... 99
Acknowledgements ... 100
References ... 100

Chapter 5 Transforming Engineering Inspections Through Advanced UAVs 103

By Alejandro Murcia Crespo, Rafael Caballero González, Miguel Ángel Trujillo Soto, Francisco Javier Pérez Grau, Antidio Viguria, Ambar Garofano Soldado, José Ramiro Martínez de Dios and Aníbal Ollero Baturone

5.1 AEROX – Elevating Inspection to New Heights ... 105
5.2 AERO-CAM – Clear Visions From Above ... 111
5.2.1 General Visual Inspection ... 115
5.2.2 Surveying Target Installation ... 116
5.2.3 Specific Visual Inspection ... 116
5.3 VIAD-DRONE – Precision Interaction ... 116
5.4 TTDRONE – Navigating the Subsurface with Expertise ... 120
5.4.1 General Tunnel Inspection ... 123
5.4.2 Local Tunnel Inspection ... 124

Chapter 6 Visualizing the End-to-End Workflow of Digital Inspections Through the Intelligence and Visualization Portal 127

By Marianna Corinaldesi, Arnaud Andrianavalomahefa, Safidy Ratsimbazafy, Patrick Ursolino, Elena Chiggiato and Luc Stakenborg

6.1 The PILOTING Platform ... 128
6.2 The Intelligence and Visualization Portal ... 130
6.3 Design and Implementation ... 132
6.4 Workflow ... 134
6.4.1 Create the Asset Digital Twin ... 137

6.4.1.1 Point clouds uploader & viewer 138
6.4.1.2 AssetBuilder tool 140
6.4.2 Define Inspection Locations 141
6.4.3 Define an Inspection Plan 142
6.4.3.1 Adding inspection plan information 143
6.4.3.2 Establishing inspection tasks 143
6.4.3.3 Selecting the robotic system 145
6.4.4 Data Exchange Between IVP and RCS 145
6.4.5 Data Analytics Tools 146
6.4.5.1 Inspection data in 3D context 147
6.4.5.2 Image annotation & AI annotations 151
6.4.5.3 File management 154
6.4.5.4 IoT dashboard 155
6.4.5.5 KPI dashboard 156
6.4.6 Reporting Tools 158
6.4.6.1 Report editor 158
6.4.6.2 PDF report generation 159
6.4.6.3 Pre-defined templates 161
6.5 Conclusion 162
Acknowledgements 163
References 163

Chapter 7 Autonomous Navigation for Inspection and Maintenance Ground Robotics **165**

By Peter Ørnulf Ivarsen, Julian Keller, Jan Sramota, Victor Reijgwart, Martin Eek Gerhardsen, Cesar Cadena, Henrik Lundkvist and Richard J.D. Moore

7.1 Introduction 165
7.2 Localization Algorithms 166
7.2.1 GNSS and Radio-Based Localization 167
7.2.1.1 Radio-based localization in GNSS denied environments 167
7.2.2 Sensor Fusion Based Localization 168
7.2.2.1 Simultaneous Localization and Mapping (SLAM) 168
7.3 Navigation Algorithms 169
7.4 Localization and Mapping of Ground Robots in PILOTING 169
7.4.1 Refinery Navigation 170
7.4.1.1 Localization and mapping 170
7.4.1.2 Motion planning and control 171

7.4.1.3 Use case: Autonomous valve manipulation 172
7.4.2 Tunnel Navigation 174
7.4.2.1 Challenges for localization and mapping in tunnels 174
7.4.2.2 PILOTING solution 174
7.4.2.3 Map creation 177
7.4.2.4 Performance evaluation 178
7.4.2.5 Use case: Autonomous inspection 179
References 180

Chapter 8 Application of Intelligent Aerial Robots to the Inspection and Maintenance of Electrical Power Lines 183

By Anibal Ollero, Alejandro Suarez, Juan Manuel Marredo, Giovanni Cioffi, Robert Pìnièka, Goran Vasiljević, Viet Duong Hoang, Michele Marolla, Jiaxu Xing, Martin Saska, Stjpan Bogdan, Emad Ebeid, Fabio Ruggiero, J. Ramiro Martínez-de Dios, Davide Scaramuzza, Vincenzo Lippiello and Antidio Viguria

8.1 Introduction 184
8.2 Intelligent Systems for Long-Range Inspection 185
8.2.1 Perception Aware Model Predictive Control for Inspection of Power Lines 185
8.2.1.1 System design and validation 186
8.2.1.2 Line detection and tracking 186
8.2.1.3 Validation 187
8.2.2 Long-Range Powerline Mapping 187
8.3 Aerial Robotic Manipulation for Maintenance 189
8.3.1 High Payload Light Arm for Aerial Manipulation Tasks 190
8.3.2 Linear Actuator Platform and Dual Arm Platform 191
8.3.3 The Main Local Manipulation Platform 193
8.3.4 Magnetic Gripper with Charging Function for Drones on Power Lines 195
8.4 Human-Machine Interfaces 197
8.5 Formation Control of Safety Aerial Co-Workers 199
8.6 Conclusions and Future Work 201
Acknowledgements 202
References 202

Chapter 9 Promoting the Use of Robotics in the Inspection and Maintenance of Offshore Wind **207**

By Andry Maykol Pinto, Aníbal Matos, João V. Amorim Marques, Daniel Filipe Campos, Maria Inês Pereira, Rafael Claro, Eeva Mikola, João Formiga, Mohammed El Mobachi, Jaap-Jan Stoker, Jonathan Prevosto, Shashank Govindaraj, David Ribas, Pere Ridao and Luca Aceto

9.1 Introduction 208
9.2 Barriers to the Adoption of Technology 210
9.2.1 Scientific and Technical Challenges 210
9.2.2 Legal and Regulatory Challenges 211
9.2.3 Certification and Trust in Robotic Technologies 212
9.3 The ATLANTIS Test Centre 213
9.3.1 Coastal Testbed 214
9.3.2 Offshore Testbed 215
9.3.3 Supporting Infrastructures 216
9.3.3.1 Infrastructures 216
9.3.3.2 Supervisory control centre 217
9.3.3.3 Mar Profundo vessel 219
9.3.4 Services and Supported Testing 219
9.4 Developments to Enable Robotic IMR of Offshore Wind Farms 222
9.4.1 Robotic Technologies 222
9.4.1.1 Unmanned Aerial Vehicles (UAV) 223
9.4.1.2 Autonomous surface vehicles (ASV) 225
9.4.1.3 Autonomous Underwater Vehicles (AUV) 227
9.4.1.4 Remotely operated vehicles (ROV) 231
9.4.2 Robotic Tools: Underwater Imaging 232
9.4.3 Digital Tools 234
9.4.3.1 Supervisory control centre (SCC) 234
9.4.3.2 Octopus 235
9.4.3.3 Tools for predictive maintenance 237
9.5 Validation of the Robotic Technologies 238
9.5.1 Coastal Testbed Tests 239
9.5.2 Offshore Testbed Tests 241
9.6 Conclusion 242
Acknowledgements 243
References 244

Chapter 10 From Reality Capture to Decision-Making 247

By José Solís Hernández, Paula López Arévalo, Noemi Jiménez Redondo, Francisco Javier Pérez Grau, Anibal Ollero Baturone and Balázs Némedi

10.1 Introduction 248
10.1.1 OMICRON's Concept 249
10.1.2 From Reality Capture to Decision Making 250
10.1.2.1 UAVs inspection technologies 251
10.1.2.2 Digital twin and decision support tool 258
10.1.3 Conclusion 267
Acknowledgements 268
References 269

Chapter 11 Robotics-Enabled Roadwork Maintenance and Upgrading 271

By Nikolaos Bakalos, Iason Katsamenis, Eftychios Protopapadakis, Carlos Martín-Portugués Montoliu, Yannis Handanos, Franziska Schmidt, Olov Andersson, Helen Oleynikova, Miquel Cantero, Ilias Gkotsis, Elena Avatangelou, Nikos Frangakis, Dimitrios Bilionis, Dimitris Vamvatsikos and Maruša Benkic

11.1 Introduction 272
11.1.1 Previous Approaches 273
11.1.1.1 The HERON contribution 274
11.2 Application Scenarios 276
11.2.1 Demonstration Sites 276
11.2.1.1 Greek demo site: Olympia Odos 276
11.2.1.2 Spanish demo site: ACCIONA 277
11.2.1.3 French demo site: Transpolis 279
11.2.2 Demonstration Scenarios 280
11.3 Computer Vision Toolkit 281
11.3.1 Performance Evaluation 282
11.4 Robotics Platform 284
11.4.1 UGV Interface 284
11.4.2 High-Level Planning 285
11.4.3 Localization 285
11.4.4 Robotic System Monitor 285
11.4.5 Rover Controller 286
11.4.6 Arm Controller 286
11.4.7 Secondary Elements Controller 286
11.4.8 Grasp Synthesis 286
11.4.9 Motion Planning 287

11.4.10 Force-Based Interaction 287
11.5 Visualisations & Improved Situational Awareness 287
11.5.1 IMS & DSS 288
11.5.2 COP 289
11.5.3 AR App 290
11.5.4 Middleware and Data Fusion 290
11.6 Conclusions 291
Acknowledgments 292
References 292

Epilogue **295**

Index **297**

About the Editor **301**

Contributing Authors **303**

Preface

Critical infrastructures are currently facing numerous challenges relating to their structural integrity being exposed to climate change, increased usage (e.g. traffic in transport infrastructures) and material ageing that are affecting and increasing risks over their ability to maintain their intended operational purpose. It is highly recognized that inspection and maintenance activities need to be enhanced and involve more precise and effective maintenance planning and targeted repairs, keeping high safety standards for personnel and increased quality.

The recent hype in digital technologies seriously supports digitalization of assets, easier and more efficient modelling and analysis of defects and thus support the infrastructures' resiliency, from data capturing, up to defects modelling, maintenance and recovery in case of more serious issues.

Robotics and automation technologies supported by other digital technologies can today contribute to various aspects of the inspection and maintenance procedures offering more precise, more effective, safer (for both infrastructure and workers) and lower cost operations. This book provides an extensive review of recent R&D efforts towards inspection and maintenance of infrastructures including robotics and automation as well as other supportive digital solutions, including Artificial Intelligence, advanced visualization, reporting and decision support etc. An overview of each technology is provided below:

Chapter 1: The book starts with this introductory chapter that defines the world of inspection and maintenance activities as well as the main industrial challenges that define the baselines of inspection and maintenance (I&M) activities, bottlenecks and robotics/automation value.

Chapter 2: This chapter presents the Inspection and Maintenance Platform of the PILOTING project directed on the critical needs for modernizing traditional I&M

methods through the adoption of remote and automated procedures. This comprises three core components: the General Robot Control Station (gRCS), the Data Distribution & Harmonization Layer (DDHL), and the Data Management System (DMS).

Chapter 3: In this chapter we present efforts on convolutional and transformer-based deep learning neural networks for the automated detection of corrosion on pipes and inside of vessels in the Oil and Gas industry. Augmentation methods such as CycleGans are also included, while experiments highlight the value of transformers outperforming convolutional neural networks both in pipes and vessels and more efficient detection of corrosion on pipes and inside vessels with high Intersection over Union (IoU).

Chapter 4: In mind of the continuous deformation assessment of motorway tunnels, this chapter presents an innovative approach for using laser scanner technologies to capture the 3D shape of assets that can be later compared to future instances to identify any deformation at millimeter accuracy. This approach is proving highly convenient and efficient as a fully non-destructive (NDT) approach.

Chapter 5: This chapter presents recent research outcomes of engineering inspections using UAVs and other autonomous aerial vehicles. The focus converges on the UAV components, shedding light on four exceptional aerial platforms: AeroX, AERO-CAM, VIAD-DRONE, and TTDRONE. Each of these UAVs has been carefully engineered with a specific purpose, leveraging state-of-the-art technologies to address the unique demands of inspections in refineries, bridges, viaducts, and tunnels.

Chapter 6: Following recent developments, this chapter presents an innovative approach for the end-to-end workflow of digital inspections through the Intelligence and Visualization Portal (IVP).

Chapter 7: This chapter presents recent research results in the frame of Autonomous Navigation for Inspection and Maintenance Ground Robotics including localization and navigation algorithms, as well as Localization and Mapping of Ground Robots.

Chapter 8: Complementing the inspection and maintenance efforts and industrial domains, this chapter focuses on the Application of Intelligent Aerial Robots to the Inspection and Maintenance of Electrical Power Lines. This research is based on the need for cost-effective solutions for powerline mapping and inspection that has motivated the development of aerial robots based on unmanned autonomous helicopters, Vertical Take-Off and Landing (VTOL), or multi-rotors.

Chapter 9: This chapter presents the use of Robotics in the Inspection and Maintenance of Offshore Wind as another highly challenging environment where autonomous robotics systems and digital transformations are proving high value.

Chapter 10: This chapter discusses the need to move from reality capture to decision-making in the context of inspection and maintenance. The chapter presents an Intelligent Road Management Platform founded on an array of innovative technologies, aimed at enhancing the construction, maintenance, renovation, and rehabilitation of the European road network. This includes the entire asset management chain, focusing on modular bridge construction, inspection digitization, predictive maintenance, and automated execution of preservation activities.

Chapter 11: Closing, this chapter presents recent research efforts and results on a Robotics-enabled Roadwork Maintenance and Upgrading approach and tools. This includes the implementation of a road infrastructure blueprint developed including advanced engineering solutions for interconnecting and facilitating seamless transitions between different transportation modes in the event of severe disruptions affecting one mode of transportation.

The target audience of this book has identified and includes the following:

1. **Infrastructure Engineers and Managers:** Personnel responsible for the design, construction, and maintenance of critical infrastructure such as bridges, roads, dams, and power plants. They will benefit from the insights into advanced inspection and maintenance approaches.
2. **Maintenance and Operations Personnel:** Technicians, field engineers, and workers involved in the day-to-day maintenance and operation of critical infrastructures will find valuable information on how new technologies can improve their work and safety keeping high quality inspection standards.
3. **Research and Academia:** Scholars and researchers in the fields of civil engineering, robotics, automation, and digital technologies may use this book as a reference for their studies and research projects.
4. **Government and Regulatory Agencies:** Personnel working in government agencies responsible for setting and enforcing safety and maintenance standards for critical infrastructures can benefit from understanding the latest advancements in inspection and maintenance.
5. **Consultants and Industry Experts:** Professionals offering consultancy services to infrastructure projects and companies looking to stay updated on the latest trends and technologies in infrastructure maintenance.
6. **Technology Providers:** Companies and individuals involved in the development and supply of robotics, automation, artificial intelligence, and digital solutions for infrastructure inspection and maintenance will find insights into market trends and potential applications.

7. **Investors and Stakeholders:** Investors considering investments in infrastructure projects or technology startups in this field may use this book to gain a better understanding of the industry and potential opportunities.
8. **General Public:** Individuals concerned about the safety and resilience of critical infrastructure in their communities and interested in the role of technology in ensuring infrastructure reliability.

DOI: 10.1561/9781638282839.ch1

Chapter 1

Introduction to Inspection and Maintenance with Robotics

By Konstantinos Loupos

1.1 Introduction

Infrastructures are ageing. This statement alone creates discussions and concerns on the types of infrastructures considered aged, the involved materials as well as their ageing effects, criticality, severity and required interventions. Ageing may include the modification or alteration of the original (and intended) structural behavior of any infrastructure that can be attributed to material deterioration (in the form of degradation) or other aspects of lifecycle engineering. The generic and overall issue of infrastructure ageing is also related to needs for premature maintenance activities or event replacement of the actual infrastructure even (or much before) the expiry of its design life which ultimately poses strong economic, environmental and social impacts. The particular material deteriorations and therefore alteration of the intended purpose of the infrastructure that can pose even safety threats, may be due to a series of reasons that can or cannot be controlled by human interventions, preferably before the damage has occurred. A long list of such causes includes earthquakes or other geological, load related (continuous or instant load that the structure cannot withstand), weather related (severe rain, snow, or even climate change effects), or other not-performed maintenance over lifecycle issues (corrosion, water leakage, spalling, delamination etc.) [1].

The above facts constitute major challenges in keeping an aged infrastructure in working, safe and its originally-planned purpose (design purpose). In mind of civil inspection of critical infrastructures, there is a long list of imperative requirements to ensure the infrastructures' resilience. Critical infrastructures are the cornerstone of modern society, providing indispensable services and support for our daily lives. They encompass a wide range, from bridges and highways to dams and power plants, playing a central role in ensuring the seamless functioning of transportation, energy distribution, and public amenities. Nevertheless, the reliability and safety of these infrastructures face constant challenges from natural forces, environmental influences, and the effects of aging. To preserve the integrity and longevity of critical infrastructures, the utmost significance lies in civil inspection and maintenance.

The inspection of structures within the field of civil engineering entails a thorough assessment of their condition, encompassing both the materials they are composed of and their overall or partial state. Numerous parameters contribute to determining the structural integrity of a building, which, in turn, may necessitate diverse actions and subsequent measures to mitigate or minimize risks and their potential impacts on safety, costs, social aspects, and operational efficiency, among others. The evaluation of civil structures' integrity holds paramount importance as it allows for the determination of their reliability levels under existing or anticipated loads, whether physical or otherwise, and the corresponding level of response required. In modern times, legal requirements impose inspection and maintenance obligations on various critical structures, including buildings, bridges, tunnels, factories, ports, power plants, and more. Given the extensive operational lifespan of such structures, there is a growing demand for a versatile system that can adapt to diverse operational needs and accommodate different structure types with varying monitoring requirements. Today, the practice of structural inspection is widely acknowledged as a means of diagnosing the health of a structure, taking into account the composition of its materials and assessing its overall or partial condition. This comprehensive evaluation encompasses a range of parameters that collectively contribute to defining the structure's integrity. Depending on the outcomes of these inspections, specific actions and measures must be undertaken to safeguard against potential risks and adverse effects related to safety, financial costs, societal impact, and operational efficiency. The significance of assessing the structural soundness of civil constructions cannot be overstated, as it allows engineers to ascertain the structure's ability to withstand present and future loads, both physical and non-physical in nature. Compliance with legal mandates further reinforces the need for periodic inspections and maintenance of critical structures, such as buildings, bridges, tunnels, factories, ports, and power plants, among others. Given the prolonged operational lifespans of these diverse structures, there is a pressing need for a flexible and adaptive monitoring system that can cater to their unique operational requirements and varying structural characteristics. This adaptive approach ensures that different types of structures receive the specific attention they require, thereby enhancing overall safety and efficiency across a wide range of infrastructure projects [3].

The following paragraph includes an analysis as an overview of the civil infrastructures' requirements that are needed per different application domain and original structure purpose [4–6].

1.1.1 Inspection and Maintenance Challenges

The ultimate and one of the fundamental requirements of civil protection is the **regular and systematic inspections** of the infrastructure. This implies the establishment of a well-structured and consistent inspection approach. To ensure the resiliency of infrastructures (including critical), it is essential to perform periodic assessments carried out by highly qualified engineers and inspectors. These assessments play a critical role in detecting issues, evaluating the current condition of infrastructure assets, and make predictions regarding their future performance and possible events escalation. The frequency of such inspections depends on various considerations and conditions and strongly depend on the infrastructure nature and purpose as well as the extent of exposure to environmental or other stressors. In parallel some critical infrastructures may necessitate more frequent inspections due to their high-risk nature, others with relatively lower susceptibility to deterioration may undergo inspections at longer intervals. The objective is to achieve a good balance between ensuring thorough evaluations and minimizing disruption to the regular operation of these crucial facilities. Disruptions for example to motorway tunnels could impose financial, safety and comfort issues. The goal of these meticulous and regular inspections is to keep the resilience, functionality, and longevity of critical infrastructures. By proactively addressing issues, investing in timely maintenance, and adhering to industry standards and safety regulations, these inspections play a pivotal role in safeguarding society, the environment, and the economy from the potential consequences of infrastructural failures. Therefore, a robust and well-executed inspection regime is an essential pillar of modern civil engineering practices, contributing significantly to the sustainable development and wellbeing of communities reliant on these vital infrastructural assets.

During rigorous inspections, highly trained engineers and qualified inspectors analyze the structural elements, components, and systems that constitute the critical infrastructure. Advanced measurement tools, non-destructive testing techniques, and monitoring equipment are employed to gather a complete dataset, ensuring that every aspect of the infrastructure's health is examined with adequate accuracy and precision. The data collected during assessments is subjected to in-depth analysis, which aids in identifying weaknesses, fatigue, material degradation, or other issues that might jeopardize the integrity and optimal performance of the infrastructure. By pinpointing these vulnerabilities at an early stage, proactive measures can be undertaken to address the identified concerns and prevent potential catastrophic failures or costly damages. On top, such inspections go beyond just detecting existing problems, as they also serve as a good basis for predictive analyses. By assessing the existing (today) state of the infrastructure, inspectors can make informed forecasts (Artificial intelligence based) regarding its future performance and potential deterioration signs. Information collected by such inspections is crucial for decision-making processes, asset management strategies, and prioritization of maintenance activities for each asset. Based on inspection findings and predictive analysis, inspection teams, safety personnel and engineering teams can determine the appropriate course of action, ranging from routine maintenance tasks to complex repairs and structural enhancements that may or not require the total shutdown of the infrastructure.

Workers and industrial personnel safety is considered additionally of critical importance as inspection activities are most of the times performed at dangerous environments. Typical examples of roadways and motorways include heavy live traffic while requires workers to be working at extreme heights, close to passing-by vehicles, under dust and moisture conditions highly risking their lives and seriously deteriorating their work life quality. Working in such risky environments poses numerous safety challenges. Workers face the constant danger of accidents and injuries due to vehicle collisions, falls from heights, exposure to harmful substances, and adverse weather conditions. The urgency of the inspection tasks may leave little room for error, and workers may be compelled to take significant risks to meet quality and efficiency standards. Other issues with safety levels include the actual inspection processes that in order to be performed as the quality and efficiency standards required they have to on-top risk human lives, as closing a section or shutting down an asset might be impossible or posing huge financial matters. In any case, however, human life and work conditions should be strongly respected and appropriate measures should be taken to combine work efficiency and high safety.

In the context of civil engineering and related structure, the establishment of a **regular and well-organized inspection procedures** is regarded as foundational. This requirement serves as the foundation of ensuring the ongoing safety, functionality, and longevity of critical infrastructures that underpin modern society (resilience). Such infrastructures compose a wide spectrum, ranging from towering bridges and intricate highway systems to dams and power plants, each playing a critical role in facilitating the smooth operation of services and functions. Execution of periodic structural assessments is a pivotal aspect of such an inspection approach. Highly skilled and qualified engineers and inspectors undertake these evaluations, leaving no stone unturned in an attempt to identify potential issues that may compromise the structural integrity and optimal performance of each asset or system component. A comprehensive assessment encompasses an in-depth analysis of various components, materials, and systems that constitute these structures. The data collected through these evaluations acts as an essential diagnostic tool, providing valuable insights into the health and stability of the infrastructure under scrutiny. On top, periodic assessments facilitate the formulation of predictive analyses as described above. This predictive capability allows for precise planning of maintenance activities, predicting potential deterioration patterns, and optimizing resource allocation.

The frequency of conducting inspections varies based on several key factors, each of which plays a crucial role in the decision-making process. The type of infrastructure in question, age, and the extent of exposure to stressors are among the significant determinants in setting the inspection intervals. For instance, highly critical and heavily utilized structures, such as major bridges and urban roadways, may necessitate more frequent inspections to mitigate potential risks and promptly address any emerging issues.

When talking about critical infrastructures, **compliance with high and demanding safety standards and regulations** seems imperative. It is considered as the ultimate steps towards sneering safety and reliability of the structures. To achieve this, inspection protocols must be carefully designed and adhered to align seamlessly with both national and international safety standards and regulations based on actual requirements and necessities. This alignment guarantees that every safety-critical aspect of the infrastructure undergoes a comprehensive and thorough examination, leaving no room for compromises that may in-turn risk the infrastructure integrity. Adherence to structural inspection standards is an important aspect of inspection processes. Engineers and inspectors work persistently to carefully inspect the structural components and systems, aiming to ensure that they satisfy the requirements set forth by established standards and regulations. Examinations include load capacity limits that constitutes another crucial element in the inspection process. Infrastructures bear the weight of varying loads and stresses during their operational lifespan. Evaluation of their load-bearing capacities is paramount to ascertain that they can safely handle the imposed loads without succumbing to undue strain or stress. Compliance with the related standard is critical during this process also. Standards and regulations additionally consider special needs and requirements for example over natural disasters that ranks among the top priorities in civil inspection. In regions prone to seismic activity, flooding, or hurricanes, infrastructures must be able to withstand the formidable forces unleashed by such events. As part of the inspection process, engineers evaluate the infrastructure's ability to endure and recover from the impact of these natural calamities, devising measures to enhance its resilience and minimize the potential for severe damage or disruption.

In the process of assessing the civil condition of critical infrastructures, various **non-destructive testing (NDT) techniques** need to be also deployed. This includes various cutting-edge methodologies and technologies that include ultrasonic testing, radiography, acoustic emission testing and other x-ray (penetrative) technologies that enable inspectors to see below the surface, peering into the inner workings of the structures without causing any surface or deeper harm. Embracing these advanced techniques has revolutionized the inspection landscape, empowering engineers and inspectors to capture critical data that would otherwise elude traditional visual inspections. Other techniques for NDT include high-frequency sound waves, that allow inspectors to penetrate the infrastructure, revealing the tiniest flaws, cracks, or voids that might be hiding beneath the surface. Radiography is another non-destructive testing method, while by employing X-rays or gamma rays, inspectors can similarly detect internal defects, such as weld discontinuities, corrosion, or material degradation, that would otherwise remain concealed and hidden under the structure surface or covering layer. Acoustic emission approaches forms another powerful tool in the non-destructive testing toolset. This technique is based on emission of transient acoustic waves while these waves are then detected and analyzed by sensors, allowing inspectors to pinpoint active damage mechanisms or potential impending failures.

In parallel to manual inspection approaches and normatives, the adoption of **structural health monitoring (SHM) systems** has emerged as a paramount necessity for critical infrastructures, driving a significant transformation in the way we safeguard and manage them. By seamlessly integrating automated monitoring solutions, SHM systems ensure a continuous watch (monitoring) over the structural behavior of infrastructures, leaving no room for potential vulnerabilities (deformations and other) to go overlooked or disregarded. Numerous essential parameters, including strain, stress, deflection, vibration, acceleration and other structural indicators, are recorded (and sometimes analyzed) in real-time, generating a large set of critical data that empowers engineers and maintenance teams with detailed insights. In other terms, with the given capacity for continuous monitoring, SHM systems offer a level of surveillance that surpasses intermittent inspection practices in a 24/7 approach. This enables engineers with a comprehensive understanding of the infrastructure's response to varying loads, environmental conditions, and external forces, offering valuable clues about its structural health throughout its operational life. By capturing and analyzing data at regular intervals, SHM facilitates the early detection of even minute anomalies, alerting stakeholders to any deviations from expected behavior well before they escalate into more significant issues (severe deformations and other). The advantages of real-time monitoring in SHM is huge. The ability to observe the structural performance of critical infrastructures as it happens in the present grants engineers advanced situational awareness and enables quick and precise decision-making, allowing for prompt action in response to emerging concerns while it supports adaptive and proactive approach to maintenance, where interventions are based on current conditions rather than relying on pre-scheduled routines [7].

The results of inspections frequently shed light on areas of significant concern within critical infrastructures, unveiling various issues like cracks, corrosion, or fatigue in essential structural elements. Addressing these with **timely and efficient inspection and maintenance plans** becomes seriously critical to safeguarding the integrity and longevity of the assets. Swift action with repairs and retrofitting measures is vital to manage the progression of deterioration, ensuring that the infrastructures remain robust and reliable long enough. Having inspection results (inspection data) serving as a guiding compass, engineers and maintenance teams can define and elaborate comprehensive strategies tailored to address the specific challenges and vulnerabilities identified for each infrastructure and asset (or subcomponent). By adhering to a well-structured plan, they can strategically allocate resources, prioritize tasks, and optimize the overall maintenance process, resulting in enhanced cost-effectiveness and minimized assets' downtime. Proactive measures through timely repairs not only averts potential hazards but also aids in the prevention of more extensive and costly future damages.

Retrofitting, as another essential aspect of the maintenance process, involves the addition or enhancement of specific elements to improve the performance and resilience of the infrastructure. It is especially crucial for aging structures that may have been designed under older standards or subjected to increased demands over time. Through retrofitting, engineers can reinforce critical components, fortify against new challenges, and extend the infrastructure's operational life, contributing to long-term sustainability. Having a precise plan for implementation of maintenance measures extends beyond the realm of immediate repairs and enhancements while it also encompasses the establishment of long-term monitoring and management programs. At the same time, the value of a well-maintained infrastructure extends beyond its longevity and immediate functionality which with proper maintenance practices also contribute to minimizing the overall life cycle costs associated with the infrastructure.

When talking about structural assessment of civil infrastructures, we should not forget a **careful consideration of environmental and climate factors** that pose significant influence on the overall condition and performance of infrastructures, making it overbearing for engineers and maintenance teams to proactively consider and address these, external, but still high, influences. For example, when considering coastal structures we need to cope with saltwater exposure, which can lead to accelerated corrosion and deterioration of key components. Similarly, freezing temperatures in colder regions pose a unique set of challenges, with the potential to inflict damage on roads and bridges through frost heave and other frost-related phenomena. As a result, the successful management of infrastructures' maintenance necessitates the incorporation of comprehensive environmental assessments into inspection and maintenance programs, enabling stakeholders to comprehend the impact of climatic conditions and undertake targeted measures to mitigate adverse effects. In addition to direct environmental impacts, climate change poses further challenges to critical infrastructures. Rising sea levels, extreme weather events, and shifting weather patterns requires adaptation and resilience strategies. Engineers must incorporate climate change projections into their assessments, considering the potential implications on the infrastructures' design life and performance. By factoring in climate change risks during maintenance planning, stakeholders can implement measures that bolster the infrastructures' ability to withstand future environmental stresses. Environmental assessments encompass an array of considerations beyond direct climate-related effects. These assessments evaluate various aspects, such as soil stability, seismic activity, air quality, and the impact on surrounding ecosystems. By understanding the ecological context and potential environmental impacts, maintenance teams can minimize the ecological footprint of infrastructure maintenance activities and uphold sustainable practices.

High requirements in the effectiveness of civil inspections and maintenance requires **highly effective, synergetic and cohesive collaboration** between inspection personnel, equipment, management and other involved stakeholders. These stakeholders may also include engineering experts, governmental bodies, private stakeholders, and research institutions, each contributing with their unique perspectives and expertise. By enabling and combining collective knowledge and resources, they can, as a team, establish a robust framework that enables improvement in inspection methodologies and establishment of best and common practices guaranteeing adequate levels of performance and effectiveness. This exchange of information enables advancing the state of civil inspection and maintenance, empowering stakeholders to enhance the resilience, safety, and performance of critical infrastructures. Structured knowledge sharing and data exchange form the lifeblood of this collaborative ecosystem remains important while stakeholders need to engage in open dialogues, sharing lessons learned from past incidents and case studies. This enables the identification of trends, patterns, and root causes of failures, leading to the implementation of preventive measures. By learning from historical experiences, collaborative efforts continually evolve, incorporating new insights into inspection methodologies and updating best practices. Moreover, a collaborative approach fosters a culture of continuous improvement and adaptation. As the collective knowledge base expands, stakeholders are better equipped to respond to emerging challenges and dynamic environmental factors. The collaborative effort becomes a platform for innovation, fostering the development of cutting-edge technologies and predictive maintenance solutions that ensure the proactive preservation of critical infrastructures.

Execution of optimal, effective and efficient civil inspection and maintenance requires serious investment in both technological advancements and **continuous training and development of inspection personnel**. This dual-side approach empowers inspection teams with the necessary resources, cutting-edge tools, state-of-the-art equipment, and invaluable expertise, ultimately paving the way for precise and meticulous assessments and informed decision-making throughout the maintenance process. Technological advancements in civil inspection have undergone remarkable evolution, offering a plethora of groundbreaking innovations and methodologies that augment traditional inspection practices. Embracing these technological marvels empowers inspection teams to transcend the limitations of conventional approaches and gain unprecedented insights into the condition and performance of critical infrastructures. Investing in the training and development of inspection personnel remain paramount especially for cases where high technology and state of the art equipment and approaches are used. Ensuring that inspectors are well-trained in the latest methodologies, industry standards, and best practices empowers them to carry out assessments with the highest level of proficiency and accuracy. Specialized training in the interpretation of data obtained from advanced sensors and inspection tools enables inspectors to derive meaningful insights and actionable recommendations from the gathered information. Moreover, creating a culture of continuous learning and professional growth cultivates a pool of highly skilled and adaptive inspection personnel. Engineers and inspectors can benefit from attending workshops, seminars, and certification programs that focus on emerging technologies and novel inspection approaches. This exposure to the latest developments in the field enables them to stay at the forefront of the industry, prepared to tackle new challenges and complexities as they arise.

1.2 Added Value of Robotic and Automation Systems Over Traditional Inspection Approaches

In mind of the above, extended, list of imminent requirements from various industries about the means, context and technologies for civil inspection and maintenance, we can think of robotic and automation systems as possible means to cope with them in an efficient and structured way [8–12].

Robotics and automation solutions provide strong advantages over manual inspections towards high and **improved safety of workers and environment**.

Inspection and maintenance activities can be performed with significantly reduced human-life risks as human presence is reduced and in any case human exposure is minimized. Robotics or automation solutions remove the needs for human 'touch' between the infrastructure workers especially at hard working conditions in hazardous or hard-to-reach areas, such as confined spaces, high heights, or areas with toxic substances, reducing the potential for accidents and injuries.

Robotic systems are engineered by design for high precision capabilities, allowing them to perform intricate and delicate inspection tasks with **exceptional accuracy and consistency**. Consistency is a key benefit of robotics in inspection and maintenance. Robots and other automation systems can also be programmed to follow predefined paths and perform tasks in a repeatable manner. Unlike human workers, who might exhibit variations in their performance due to different factors, robots maintain the same level of performance regardless of external and other working conditions. This consistency ensures that inspections are carried out uniformly across different areas or assets, minimizing the risk of overlooking critical issues. Human error is an inherent risk in any manual inspection or maintenance activity. Fatigue, distractions, or lapses in attention can lead to oversight or misjudgment, potentially compromising the accuracy of the assessment or the quality of the maintenance work. Thus, the risk of errors due to human factors is drastically reduced.

The combination of precision, consistency, and reduced human error leads to reliable and repeatable outcomes in inspection and maintenance tasks. With robotics, the results of inspections are highly dependable and can be replicated consistently over time. This reliability is especially crucial in critical industries such as aerospace, energy, and transportation, where even minor errors can have severe consequences.

Financial benefits of using robotics and automation for inspection and maintenance operations are still large, despite the initial investment that is outweighed. This initial cost encompass acquiring robotic systems, integrating them into existing processes, and training personnel to work alongside these machines. Despite the initial financial outlay, the long-term benefits offered by these technologies often outweigh the costs. By tradition, inspection and maintenance tasks require substantial workforces, involving skilled technicians, workers and laborers. Robots can take over many of these tasks, significantly reducing the need for human labor. This reduction in personnel requirements translates to lower payroll expenses, saving companies substantial amounts over time. Financial effects also come over improved safety that not only lower human risks but also contributes to reduced insurance costs for companies. Insurance premiums are often influenced by the safety record of a company, and implementing robotics and automation can lead to lower insurance expenses in the long run.

Robotic systems are able to work non-stop, 24/7, without rest or other vacation time, leading to very much **improved efficiency and low down-time** while at the same time they can work faster and in parallel as compared to traditional manual methods. This continuous operation ensures that inspection and maintenance activities progress at a steady pace, maximizing the utilization of available time and resources. On top, robotics, excel in speed and precision, enabling them to perform tasks much faster than humans. They can navigate through complex structures swiftly, conduct detailed inspections, and execute maintenance actions efficiently while their parallel capabilities allow multiple tasks to be carried out simultaneously (e.g. fleet of drones inspecting various sections of a vast infrastructure concurrently, accelerating the overall inspection process). Their **reduced response time** and being unaffected by environmental conditions are very useful features for scenarios where immediate action is required, such as detecting and repairing critical equipment failures or infrastructure damage, robots can be quickly dispatched to the site, minimizing response times. This swift response contributes to avoiding extended downtime and minimizing potential losses due to delays.

A major requirement of efficient inspection processes is proper **reporting and data management**. Without being highly competent in this aspect, the value of even highly precise inspection missions can significantly reduce and prove useless. Robotics and automated systems for data collection and structuring can prove beneficial regarding proper data management, storage, structuring even when talking about of vast amounts of data (e.g. civil inspection data of a 3 km length engineering structure, a motorway tunnel). Digital data management systems can be of course combined with an inspection reporting system, allowing for real-time, valuable insights of asset, equipment and infrastructure. This can be on-top combined with predictive simulation or predictive maintenance systems that can in turn anticipate issues, damage escalation and lead predictive maintenance efforts and planning. In mind of digital reporting and connectivity aspects, we need to additionally highlight the capabilities of robotics systems on **remote operation and telepresence**. These apart from keeping human safety risks low, support to conduct inspections and maintenance tasks from a safe location, even if the asset is located in a challenging or remote environment. Telepresence capabilities enable skilled technicians to guide less experienced personnel on-site, reducing travel costs and response times.

1.3 Market Overview and Positioning

The global market for inspection and maintenance with robotics has notably increased steaming from the increased (and increasing) demand for inspections in critical and hazardous environments. This rise can be attributed to the increased recent concerns for safety and efficiency in various industries including oil and gas, manufacturing, transport, avionics etc. Aims behind the even stronger adoption of inspection robotics stems from the necessity to conduct inspections in locations that pose significant risks to human workers while at the same time ensure the monitoring of critical infrastructures that will ensure their resilience and keep their intended purpose.

Despite the numerous advantages of robotics and automation systems as defined above, widespread acceptance of inspection and maintenance robots from consumers still remains a strong challenge for all actors and stakeholders. This reluctance to embrace such technologies could be attributed to concerns about the reliability and data accuracy, as well as strong fears of potential job displacement among human workers. On top, further factors that hampers the growth of this market is the sensitivity and handling capabilities of these robots. Ensuring that inspection robots can handle delicate equipment and structures without causing any further damages remains crucial and needs further efforts in the social acceptance direction. Addressing these issues would not only increase the market's growth potential but also boost trust and confidence in the deployment of these robots in critical industries. Despite the above aforementioned challenges, other factors are driving the growth of the EU and worldwide inspection and maintenance robotics market. The given and foreseen/predicted escalating focus on industrial automation as well as increased needs for optimized operational efficiency are factors that are driving industries to seek advanced solutions for inspection and maintenance operations in various industries. On top, the increasing awareness about the benefits of early detection and prevention of potential issues is encouraging industries to adopt inspection robots as a proactive approach to avoid costly downtime and safety incidents.

The inspection and maintenance with robotics market has a huge potential and is expected to largely expand in the following years, posing strong value for various industries with a series of benefits as outlined in the previous chapters. The market size for inspection robotics was valued at M$940.0 in 2020, while it is projected reaching M$13,942.5 through 2030. This sums to a CAGR of 30.9% (2021 to 2030). Regarding the market segments, the inspection robotics market is right now segmented by type, application, end user, and region. Further segmentation can also be structured into stationary and mobile robotics (by type). This should be also covering automated inspections and non-destructive ones [13, 14].

1.4 Conclusions

In this introductory chapter we have highlighted the challenges posed by ageing infrastructures and the importance of civil inspection and maintenance in ensuring the resilience and safety of critical infrastructures. The chapter emphasizes the need for periodic and thorough assessments carried out by highly qualified engineers and inspectors to detect issues and predict future performance. Safety of workers during inspections is a significant concern, especially in hazardous environments, and the article stresses the importance of ensuring worker safety while conducting inspections. Further, this section has presented the added value of robotic and automation systems over traditional manual inspection approaches. Robotic systems offer several advantages, including enhanced safety, precision, consistency, and reduced human error. They can work non-stop, 24/7, increasing efficiency and minimizing downtime. While the initial investment may be high, the long-term benefits, such as lower labor costs and insurance expenses, outweigh the costs. Furthermore, the chapter highlights the importance of data management and reporting in inspection processes. Robotics and automation systems aid in proper data structuring and management, allowing for real-time insights and predictive maintenance planning. Remote operation and telepresence capabilities enable inspections and maintenance tasks to be conducted from safe locations, reducing travel costs and response times.

Overall, the article emphasizes the critical role of civil inspection and maintenance in ensuring the safety, functionality, and longevity of critical infrastructures. It highlights the potential benefits of integrating robotic and automation systems in inspection processes to enhance efficiency, precision, and above all, workers' safety.

References

[1] K. Loupos, A. Amditis, C. Stentoumis, P. Chrobocinski, J. Victores, A. Roncaglia, S. Camarinopoulos, N. Komodakis, R. Lopez – ROBOTIC SYSTEM WITH INTELLIGENT VISION FOR TUNNEL STRUCTURAL ASSESSMENT – SYSTEM ARCHITECTURE – THE ROBO-SPECT-EC PROJECT – Third Conference on Smart Monitoring, Assessment and Rehabilitation of Civil Structures, SMAR 2015, 7–9 September 2015, Antalya, Turkey, ISBN: 978-3-905594-65-2.

[2] K. Loupos, A. Amditis, A. Doulamis, P. Chrobocinski, J. Victores, M. Wietek, P. Panetsos, A. Roncaglia, S. Camarinopoulos, V. Kallidromitis, D. Bairaktaris, N. Komodakis, R. Lopez – INTEGRATED ROBOTIC SOLUTION FOR TUNNEL STRUCTURAL EVALUATION AND CHARACTERIZATION – ROBO-SPECT EC PROJECT. International Conference on Smart Infrastructure and Construction (ICSIC), 27–29 June 2016, Robinson College, Cambridge.

[3] K. Loupos, A. Amditis, A. Tsertou, Y. Damigos, R. Gerhard, D. Rychkov, W. Wirges, V. Kalidromitis, S. Camarinopoulos, S. Lenas, V. Tsaoussidis,

A. Anastasopoulos, K. Lenz, S. Schneider, M. Hill, A. Adesiyun, B. Frankenstein – SKIN-LIKE SENSOR ENABLED BRIDGE STRUCTURAL HEALTH MONITORING SYSTEM, 8th European Workshop on Structural Health Monitoring (EWSHM), 5–8 July 2016, Bilbao, Spain.

[4] K. Loupos, A. Amditis – FIBER OPTIC SENSORS: CURRENT STATUS AND FUTURE POSSIBILITIES, Chapter: STRUCTURAL HEALTH MONITORING FIBER OPTIC SENSORS, Springer-Verlag Lecture series of Smart Sensors, Instrumentation and Measurements, ISBN: 978-3-319-42624-2. DOI: 10.1007/978-3-319-42625-9.

[5] E. Protopapadakis, C. Stentoumis, N. Doulamis, A. Doulamis, K. Makantasis, K. Loupos, G. Kopsiaftis, A. Amditis – AUTONOMOUS ROBOTIC INSPECTION IN TUNNELS, XXIII International Society for Photogrammetry and Remote Sensing (ISPRS) Congress, Prague, July 2016.

[6] K. Loupos, A. Amditis – FIBER OPTIC SENSORS: CURRENT STATUS AND FUTURE POSSIBILITIES, Chapter: STRUCTURAL HEALTH MONITORING FIBER OPTIC SENSORS, Springer-Verlag Lecture series of Smart Sensors, Instrumentation and Measurements, 3 November 2006, ISBN: 978-3-319-42624-2. DOI: 10.1007/978-3-319-42625-9.

[7] K. Loupos, A. Amditis, A. Tsertou, Y. Damigos, R. Gerhard, D. Rychkov, W. Wirges, V. Kalidromitis, S. Camarinopoulos, S. Lenas, V. Tsaoussidis, A. Anastasopoulos, K. Lenz, S. Schneider, M. Hill, A. Adesiyun, B. Frankenstein – SKIN-LIKE SENSOR ENABLED BRIDGE STRUCTURAL HEALTH MONITORING SYSTEM, 8th European Workshop on Structural Health Monitoring (EWSHM), 5–8 July 2016, Bilbao, Spain.

[8] K. Loupos, C. Stentoumis, A. Doulamis, N. Doulamis, A. Amditis, E. Protopapadakis, P. Chrobocinski, J. Victores, R. Montero, E. Menendez, C. Balaguer, R. Lopez, M. Cantero, R. Navarro, A. Roncaglia, L. Belsito, N. Komodakis, P. Singh, S. Camarinopoulos – AUTONOMOUS ROBOTIC SYSTEM FOR TUNNEL STRUCTURAL INSPECTION AND ASSESSMENT, International Journal of Intelligent Robotics and Applications (IJIRA), Springer, 11 August 2017, DOI: 10.1007/s41315-017-0031-9.

[9] E. Protopapadakis, C. Stentoumis, N. Doulamis, A. Doulamis, K. Makantasis, K. Loupos, G. Kopsiaftis, A. Amditis – AUTONOMOUS ROBOTIC INSPECTION IN TUNNELS, XXIII International Society for Photogrammetry and Remote Sensing (ISPRS), ISPRS Annals of Photogrammetry, Remote Sensing and Spatial Information Sciences, Prague, July 2016, DOI: 10.5194/isprs-annals-iii-5-167-2016t.

[10] K. Loupos, Y. Damigos, A. Tsertou, A. Amditis, S. Lenas, C. Chatziandreoglou, C. Malliou, V. Tsaoussidis, R. Gerhard, D. Rychkov, W. Wirges, B. Frankenstein, S. Camarinopoulos, V. Kalidromitis, C. Sanna, S. Maier,

A. Gordt, P. Panetsos, LIFECYCLE ANALYSIS AND ASSESSMENT IN CIVIL ENGINEERING: TOWARDS AN INTEGRATED VISION, 1st edition, Robby Caspeele, Luc Taerwe, Dan. M. Frangopol. CRC Press, IBSN: 9781138626331 – CAT# K31841.

[11] E. Sdongos, A. Tsertou, N. Georgakopoulos, K. Loupos, A. Amditis, N. Joram, F. Ellinger, B. Lindner, E. Edwan, A. Ehlerding, S. Camarinopoulos, V. Kallidromitis, D. Bairaktaris, E. Bairaktaris, J. Naundrup, N. Kerle, M. Gerke, K. Vierhuß-Schloms, M. Markus – A NOVEL & PRACTICAL APPROACH TO STRUCTURAL HEALTH MONITORING – THE RECONASS VISION LOCAL POSITIONING, SENSOR NETWORKS, SECURE COMMUNICATIONS AND REMOTE SENSING AT THE SERVICE OF STRUCTURAL MONITORING TO ASSESS CONSTRUCTION DAMAGE AND RELATED NEEDS, 2014 IEEE Workshop on Environmental, Energy, and Structural Monitoring Systems – 2014 IEEE Workshop on Environmental, Energy, and Structural Monitoring Systems, 17–18 September 2014, Naples, Italy.

[12] K. Loupos, A. Amditis, C. Stentoumis, P. Chrobocinski, J. Victores, M. Wietek, P. Panetsos, A. Roncaglia, S. Camarinopoulos, V. Kallidromitis, D. Bairaktaris, N. Komodakis, R. Lopez – ROBOTIC INTELLIGENT VISION AND CONTROL FOR TUNNEL INSPECTION AND EVALUATION – THE ROBINSPECT EC PROJECT, 2014 IEEE International Symposium on Robotic and Sensors Environments – IEEE International Symposium on Robotic and Sensors Environments, 16–18 October, 2014. Timisoara, Romania.

[13] Global Inspection and Maintenance Robot Market – Analysis and Forecast, 2020–2025, Focus on Type, Component, and End User, 2020.

[14] Inspection Robots Market by Robot Type (Stationary Robotic Arm and Mobile Robots), Testing Type (Automated Metrology and Non-destructive Inspection) and End-user (Oil & Gas, Food & Beverages, Pharmaceuticals, Electronics, and Others): Global Opportunity Analysis and Industry Forecast, 2021–2030.

DOI: 10.1561/9781638282839.ch2

Chapter 2

Integrated Inspection and Maintenance Platform for Critical Infrastructures

By Georgios Sarigiannis, Athanasios Karydis, Konstantinos Loupos, Eleni Voulgari, Aikaterini Karagiannopoulou, Konstantinos Routsis, Valantis Tsiakos, Georgios Tsimiklis, Angelos Amditis, Marco Antonio Montes Grova, Joaquín Borrego Villa, Francisco Javier Pérez Grau and Antidio Viguria

The deterioration of European infrastructure assets, exacerbated by the ongoing economic challenges within Europe and the escalating costs associated with maintenance, has necessitated the pursuit of innovative, sustainable, and cost-effective approaches to autonomous inspection and maintenance (I&M). This chapter presents a pioneering endeavour within the EU-funded project PILOTING (No. 871542), which addresses the critical need for modernizing traditional I&M methods through the adoption of remote and automated procedures. The PILOTING project introduces a versatile framework, comprising three core components: the General Robot Control Station (gRCS), the Data Distribution & Harmonization Layer (DDHL), and the Data Management System (DMS). This study provides a comprehensive exploration of this framework, elucidating its intricate design and operational intricacies. By embracing the PILOTING ecosystem, this research seeks to underscore the profound advantages it offers to the realm of infrastructure management. In a landscape where safety, efficiency, and innovation intersect, the PILOTING project illuminates a path towards a more resilient and promising future for the inspection and maintenance of critical infrastructure.

2.1 Introduction

PILOTING Inspection and Maintenance (I&M) Platform stands as a testament to the convergence of cutting-edge robotics and intelligent data management.

This chapter delves into the heart of the PILOTING project, shedding light on the dynamic landscape of infrastructural inspection and maintenance. Focusing on the I&M Platform developed as integral component of the project, we unravel the intricate web of technologies and strategies that underpin this holistic approach. As we journey through the chapter, the multifaceted dimensions of the platform's design, functionality, and benefits come to the fore.

Modern engineering endeavours are not without their share of challenges, and the inspection of complex infrastructures presents a unique set of hurdles. The demands of cost-effective data collection, seamless data transformation, and the synthesis of meaningful insights from vast amounts of data are just a few among these challenges. Moreover, the management of data heterogeneity and the assurance of information quality stand as critical imperatives in an age where data drives decision-making.

In response to these challenges, the PILOTING I&M Platform emerges as a beacon of innovation, ushering in a new era of efficiency and safety (Figure 2.1). As a cohesive amalgamation of hardware and software components, this platform is meticulously designed to elevate the quality and efficiency of inspection and maintenance activities across a spectrum of infrastructures. European refineries, bridges, viaducts, and tunnels – each finds a tailored solution within the versatile framework of the PILOTING ecosystem.

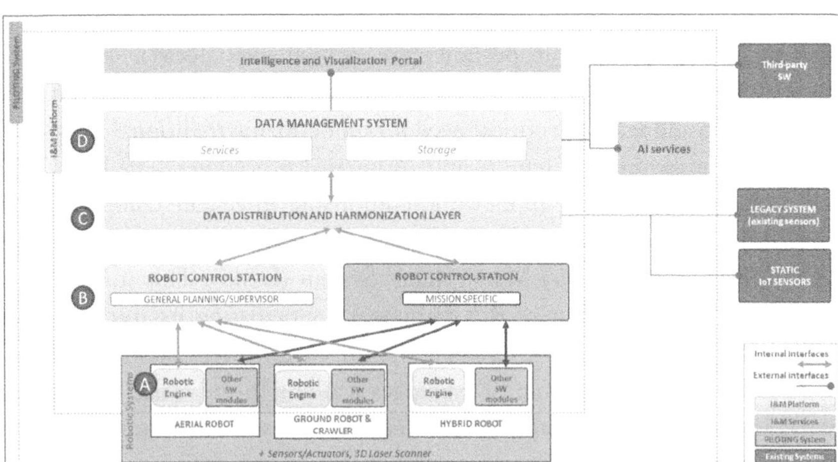

Figure 2.1. System architecture.

Central to the I&M Platform's success are three fundamental components: the General Robot Control Station (gRCS), the Data Distribution & Harmonisation Layer (DDHL), and the Data Management System (DMS). These interconnected pillars facilitate the orchestration of robotic systems, harmonization of heterogeneous data, and the secure storage and retrieval of crucial information.

- The gRCS, acting as the node of connectivity, empowers seamless communication and coordination among diverse robotic systems. It breathes life into the collaboration between Unmanned Aerial Vehicles (UAVs), Unmanned Ground Vehicles (UGVs), and the Inspection and Maintenance Platform itself.
- The DDHL takes center stage in bridging the gap between the gRCS and the Data Management System. By transforming disparate data into a standardized format, it streamlines the flow of information, ensuring a unified foundation for meaningful analysis.
- At the core of this platform, the DMS stands as the custodian of data. With a scalable architecture, it not only ensures secure storage but also facilitates authorized access, retrieval, and indexing of data accumulated throughout the inspection process.

In the following sections, we delve into each facet of the I&M Platform, exploring its architecture, capabilities, and impact on the landscape of infrastructural management. Through comprehensive analysis and insight, we seek to empower engineers and practitioners with the knowledge needed to navigate the complex realm of integrated inspection and maintenance.

2.2 A Scalable Data Management System Facilitating the Autonomous Inspection and Maintenance of Structural Health Monitoring Operations

2.2.1 Introduction

Existing civil and industrial Critical Infrastructures (CIs) have been deteriorating over the past years due to a plethora of non-neglected factors such as natural hazards and environmental destructions, and as an outcome of intense daily operational loads [1], delays and shortcomings in maintenance operations. Furthermore, the long-lasting financial crisis that affected many European countries further exacerbated the depletion of the maintenance procedures, which are mandatory to occur every 1–2 years. Traditional processes of CIs inspection have shown relatively poor performance for early warning that could prevent destructive events, since, it incorporates obsolete techniques and the physical on-site presence of dedicated personnel. Such a prerequisite is characterised as extremely costly and therefore prone to work accidents as employees expose themselves to dangerous and in some cases hard to escape environments or gaseous pollutants that can produce severe respirational impacts, as well as demolitions with many casualties.

This way, the Structural Health Monitoring (SHM) Systems have been described as a paramount importance infrastructure in the I&M, as they support obtaining real-time information from distance [2]. By and large, SHM denotes a broad interdisciplinary scientific domain that focuses on technologies and methods, which facilitates the continuous assessment and evaluation of CIs' health condition and at the same time introduces digital practices in a scalable and cost-effective manner [3]. Integrating the SHM systems into the lifecycle of the I&M strategy provides the ability to assess the structural performance in real-time, but most essentially facilitates the decision-making, formulation and prioritisation of the proper countermeasures regarding the maintenance, increasing serviceability and the assurance of safety [4]. Currently, large datasets differentiating in content and data and metadata structure are becoming available at an exponential rate, as several sensing systems, e.g., autonomous robotic systems covering the land, air and marine environments, in-situ static sensors or even crowdsource contributions [5] are utilised, promising a holistic, repetitive, secure and cost-efficient solution that can improve the efficiency and reliability of SHM systems [6]. With such a disruptive impact of digitalisation in many domains, the DMS has now become an essential part of the CIs I&M and SHM procedures [7].

Nevertheless, until nowadays, the attention has been solely given to sensing technologies, such as wireless transmission, system identification, damage detection and safety assessment. Conventional practices of such methods are described by the analysis of reports, drawings and photos or measurements retrieved by expensive equipment during on-site inspections, usually stored in local-storage personal computers (PCs). Such methods are prone to subjective errors, project-based solutions and tools for processing and analysing these data sources, and in general, slower, non-scalable and non-interoperable practices, which prohibits the data mining and storage of huge, heterogeneous and real-time data rates. Hence, it seems that there is a paucity of research addressing and improving the overall architectural optimisation for data management and maintenance [8, 9]. Previous research has posed significant technical challenges over practices and technologies that deal with processing, analysing, understanding and maintaining the collected information [10, 11], which are expressed by the five following obstacles: (i) the access to high-quality data, (ii) the lack of maintenance these observations that lead to knowledge loss over time, (iii) the absence of a secure system that could prevent any conflicts in sharing sensing information that normally CI companies and governances steward, (iv), machine-readable and standardised data access that will facilitate data transferability and comparability, culminating in (v) a lack of collective intelligence [12].

Contrariwise, the DMS has been declared as a novel element in SHM systems, as it is able to support the data mining or knowledge discovery from a vast amount of data that covers design, construction maintenance information and measures corresponding to the health state and the temporal evolution of the structure [13]. In traditional SHM systems, three essential modules are usually incorporated: the sensing device, responsible for the data collection activities, the DMS for any procedure related to the data transmission, consolidation, secure storage and maintenance, and the health assessment system that performs the appropriate analysis and extracts meaningful findings of the condition stage of the inspected infrastructures [14]. With respect to the intermediate module, two modes have been introduced so far in internet-based data distribution applications; the Browser/Server (B/S) and the Client/Server (C/S), and it can be concluded that in both cases the development of the backend-server is a prerequisite for operating any SHM system, and therefore it should be placed at the spotlight. According to the literature, the 3-tier architecture, service-oriented architecture, and micro-service architecture are the backend design concepts that are commonly adopted. Referring to the 3-tier architecture, which is the first and by far the most adopted solution, is mainly comprised of three layers: the data access layer, the business logic layer and the user interface layer, which is usually deployed in a single server and operates the user requests by various controllers that exist in the business logic layer. Such a solution can be applicable in small-scale applications and business units, however lately, it has been characterised by poor performance, hard applicability and maintenance cost. To surmount these bottlenecks, the SOA [15] has been established as a new-generation solution, which is suitable for large-scale applications. SOA proposes a composition of self-contained computing components, called services, which can operate by handling the data inputs and outputs in a common format, with the flexibility to not be tightly coupled so as to operate [16], thus giving the ability of different systems or components to be integrated, exchanging or capitalising on the information that is generated [17]. Finally, micro-service architectural designs are coming to further expand SOA principles by offering a more distributed extension, storage, and computing pressure balance by distributing function modules to multiple service instances [9].

Relying on these research findings and our previous experience, this book chapter aims to present a scalable, secure and interoperable DMS that was developed under the framework of the PILOTING project. We draw our attention to the development of a relational data model that will enable the integration of any type of field device (e.g. both robotic vehicles and in-situ sensors) and successfully store and index any data asset created in the whole I&M lifecycle. For this matter, we capitalised on several open-access technologies and the most known standards, avoiding providing a toilsome and hard-to-deploy solution. With respect to the last and due to the vision of the PILOTING project to sincerely advocate the Open Science dimensions, we decided to disseminate through the Creating Commons Attribution 4.0 International (CC BY 4.0) and the Apache 2.0 open-accessed licences, the DMS Entity-Relationship-Diagram (ERD) data model and the API (both source code and handbook). The above-mentioned are provided as web sources at the end of this chapter.

The rest of this chapter is structured as follows. Section 2.2.2 outlines the general overview of the architectural design of DMS, introducing the main modules and functionalities that satisfy. Based on the preliminary definitions, the DMS Customised data model is presented, acting as a backbone and consolidation of the six defined modules of DMS. Subsequently, the data model of the SensorThings API (STA) standard is introduced. Section 2.2.3 analyses the functionalities and business logic of the main six modules. Section 2.2.5 provides additional information regarding the implementation and hosting of all the aforementioned technological components and finally warrants high-level performance through the evaluation of certain functionalities. Finally, conclusions are presented in Section 2.5.

2.2.2 Development Framework of the I&M PILOTING Data Management System

2.2.2.1 System architectural overview

Figure 2.2 illustrates an overview of the PILOTING DMS architectural design and the six key components that consist of, namely (1) the PostgreSQL database, (2) the Mission Component Service, (3) the Mission/Asset Centric storage service, (4) the AI metadata service, (5) the Static Sensors' storage service, and (6) the User Management and Authentication module. Thorough descriptions and their beneficial contributions to the overall architecture and the project are obtained in Section 2.2.3. Adopting the SOA architectural design, we advocated our tendency to overcome the traditional challenges [17] mainly related to the (i) absence of uniform standardisation in communication protocols, (ii) interoperability between different modules, (iii) security, privacy and trust mechanisms utilising authentication, encryption and access control, [18] and (iv) scalability of resource management and resilience [19]. As a result, we presume that the DMS RESTful API Gateway facilitates the quick adaptation between various applications, without significant impacts on user demands, and thus introduces a solution with reusable technology and the agility to integrate new devices and modules in a "plug-and-play" manner.

We will start the analysis, by presenting the foundation of this system, which is the specification of the business requirements and the subsequent three operational

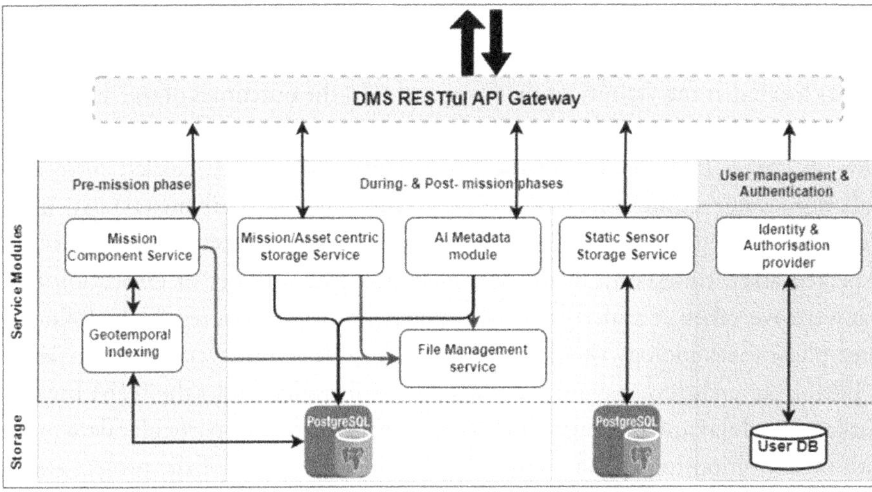

Figure 2.2. DMS Software architectural design.

phases of the I&M lifecycle, as both of them were responsible for the definition and orchestration of the set of logical modules, and therefore internal and external interfaces.

- **Pre-mission phase:** This particular phase incorporates the earliest stages of the inspection planning process, in which 3D models, inspection locations and complementary files are imported/created in the visualisation interface and are available in the DMS. With these data sources available, entities related to the inspection plan, inspection tasks, files, the assigned personnel, robotic system and payloads can be defined along with information about inspection locations and orders of mission tasks' execution.
- **During-mission phase:** Describes processes and data created on-site, where the respective users leverage on data created in the pre-mission phase, so as to define the mission and the corresponding sub-tasks that are scheduled to be implemented by the assigned robotic system and payload(s).
- **Post-mission phase:** Finally, the post-mission phase is the closure of the inspection planning operation, where all the collected information is forwarded to the DMS, allowing the generation of summarised reports and statistical analysis based on the collected findings.

2.2.2.2 A three-phase methodology for the design of the DMS data model

The ultimate ambition of the DMS is to cope with the extremely heterogeneous nature of the inspection and sensing data that are collected during the whole lifecycle of the I&M. Thus, from the beginning of the architectural design of the DMS, a not neglected demand was to develop a scalable data storage solution, which will enable the integration of nine different robotic vehicles, static accelerometer in-situ sensors located at the vicinity of the inspected CIs, the outcomes of the data-driven Artificial Intelligence (AI) models that analyse the collected information and thus showcases areas that are characterised by various defects (i.e., cracks, corrosions, spallation, water sediments, etc.), and the complementary static (meta-) data information describing the details arising from the inspection/mission procedures (e.g., inspected asset, timestamp of mission execution, the number of inspection tasks that were overtaken at a specific order and location, etc.). As a result, the following three-phase methodology was adopted in order to design the versatile Data Model of DMS. Nevertheless, having the intention of introducing to the I&M business market a modular and data-agnostic solution, we created this particular data model with the ambition to expand its usability beyond the extent of the project and be able to be incorporated with ease in any related structural monitoring use case, and thus to ensure its future sustainability.

Starting the analysis with the First Phase, we semantically defined the entities of the PILOTING ecosystem, which are the inspection pilot sites, denoted as assets, and the different procedures that were important to data providers and asset owners. During the Second Phase, we managed to identify the relationships between entities, focusing on information that must pre-exist for entities to be semantically accurate, hierarchies between entities, especially physical assets, and data needed to form an inspection plan. Finally, in the third phase, we collected the outcomes from the previous two stages and thus created the data model, taking into consideration any possible integration with third parties. The outcome of the aforementioned procedure is illustrated in Figure 2.3.

2.2.2.3 The standardised OGC-Compliant SensorThings data model

An exemption to the above-mentioned data model was identified as a necessary step, due to the different nature of the data collection process, when static in-situ sensors are deployed. In general, the rationale behind the design of the DMS API is indissolubly related to the inspection planning time schedules. This condition denotes that the data-gathering process is by default a time-restricted procedure, as it occurs until the completion of the specific mission task. To overcome this constraint and permit continuous data collection from the static sensors, we introduced the SensorThings API [21] data model from OGC [20]. The SensorThings API is an OGC standard that provides an open and unified framework to interconnect sensing devices with data and applications. It follows REST principles, JSON encoding, the OASIS OData protocol and URL conventions. The foundation of the SensorThings API is based on the ISO 19156, ISO/OGC Observations

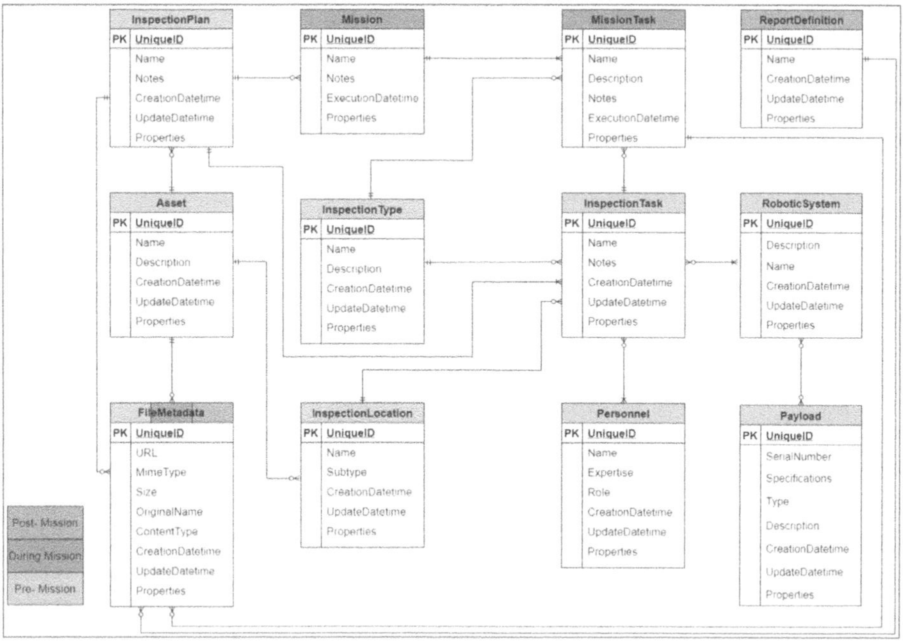

Figure 2.3. The conceptual design of the DMS data model alongside the class names and their relations. Attributes and parameters are not presented, in terms of simplifying the information given.

and Measurements, which defines a conceptual model for observations and features incorporated in data collecting observations [22].

SensorThings API fulfils two data models, the Sensing, offering CRUD (create, read, update, delete) capabilities and the Tasking, which is responsible for the parameterisation of IoT devices (sensors and actuators) [23]. According to the identified needs of the PILOTING project, we decided to employ only the Sensing data model (Figure 2.4). Subsequent relations between the entities of the DMS custom data model and the SensorThings API are provided, thereafter.

- **Sensor:** description of an IoT sensor that provides measurements of the exact location, and details related to the sensing device (e.g. image, sensor's manual, etc.).
- **Thing:** real-world thing, referring to the Asset Parent (i.e., Refinery, Viaduct or Tunnel). The unique identification index (UUID) will be identical to the Asset entity of the DMS.
- **Location:** Thing/Asset Parent location.
- **DataStream:** links a Thing with a Sensor that is measuring an Observed Property to provide an entry point for a time series.
- **ObservedProperty:** The observed physical property

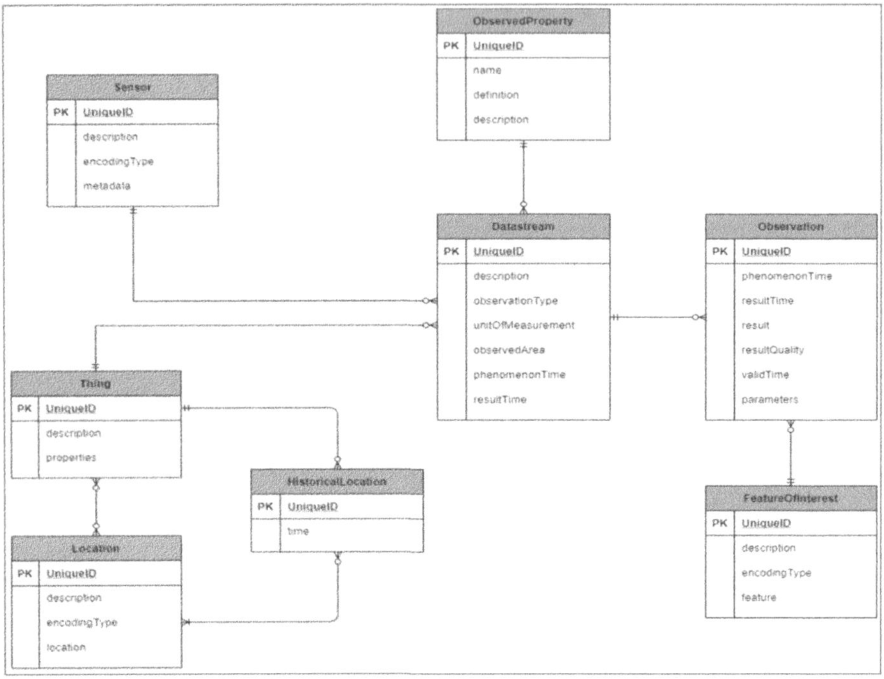

Figure 2.4. The OGC-SensorThings API Data Model (Sensing) with the corresponding class names and their relations [21].

- **Observation:** A single measurement value
- **FeatureOfInterest:** the Object/Asset Child on which the measurement was performed.

2.2.3 The Data Management System Components

2.2.3.1 PostgreSQL relational database for data management (RDBMS)

Given the need for a scalable storage solution that can support the provision of both spatial and non-spatial data observations with occasionally variant time distributions and unknown data volume and the need to efficiently respond to spatiotemporal data queries from multiple tasks over the inspected asset [22], the PostgreSQL relational database was used for both DMS and SensorThings APIs with the PostGIS extension [24], satisfying the persistent storage layer. All the data management operations performed over both APIs are based on object-relation mapping (ORM), which essentially reduces the complexity of accessing the database [17]. In general, PostgreSQL is an open-source relational database server, while PostGIS is an open-source spatial database extension, offering various geospatial filtering operations. The structure of the database can consist of several schemas, and host relational

Table 2.1. The File Management Service endpoints to retrieve the collected data/metadata.

I&M phases	DMS end-points
Pre-mission phase	*https://isense-piloting.iccs.gr/apis/**Asset**/{UUID}/FileMetadata/* *https://isense-piloting.iccs.gr/apis/**InspectionPlan**/{UUID}/FileMetadata/* *https://isense-piloting.iccs.gr/apis/**InspectionTask**/{UUID}/FileMetadata/*
During-/Post-mission phase	*https://isense-piloting.iccs.gr/apis/**Mission**/{UUID}/FileMetadata/* *https://isense-piloting.iccs.gr/apis/**MissionTask**/{UUID}/FileMetadata/*
Post-mission phase	*https://isense-piloting.iccs.gr/apis/**ReportDefinition**/{UUID}/FileMetadata/*

tables with various metadata types, ready to be harvested or appended by means of different kinds of SQL statements. Additionally, the latest versions of PostgreSQL provide the advantage of introducing the JSONB data type, a direct competitor to NoSQL technologies, as it gives the flexibility to incorporate data in an optimised data format, including a variety of properties, such as storing highly nested data whose structure may alter over time, JSON arrays, concatenated JSON objects, etc., giving at the end a more dynamic and advanced data query solution [25].

2.2.3.2 File management service

As can be seen in Figures 2.2 and 2.3, the File Management functionality has a horizontal role in the DMS, in conjunction with other functionalities. It is responsible for storing and handling all the file-based resources that are created during the overall I&M procedure. The files stored in the File Management Service and always associated with the corresponding metadata and the reference to their storage location. Indicatively, a batch of images that are captured by an on-robot camera during a mission task initially stores the collected observations and the corresponding metadata in the Mission and Asset-centric storage service. Subsequently, the image is stored in the File Management Service, incorporating in the image metadata the reference path to its storage location. All this information can be retrieved by performing the following requests, (Table 2.1).

2.2.3.3 Mission component service

As it was thoroughly revealed in Section 2.2.2, the pre-mission phase involves the initial procedures of the inspection lifecycle, and these usually incorporate the activities of planning and allocation of inspection resources, e.g. the infrastructure assets, the robotic systems and the corresponding payloads, the assigned personnel and procedures that are planned to be executed in the respective locations within an asset and by a sequential order. This functionality contains every possible process or flow required to enable users to be informed and aware of the available components, assets and tools that can be utilised in the context of an inspection mission. Thus, the Mission Component Service allows the logical relation of inspection resources and the creation of detailed inspection plans and subsequent tasks, depending on periodic and specific requirements of the infrastructure maintainer and owner.

In particular, two main operations are satisfied by this service, the mission management service functionality that assists with the definition of the infrastructure-related assets, and therefore the assignment of the necessary inspection equipment and procedures. Such information is declared through the pre-mission entities of the DMS API, e.g. Asset, Inspection Type, Inspection Location, Robotic System, and Payload (see Figure 2.3). All the aforementioned entities finally synthesise the Inspection Task and Inspection Plan entities. After the execution of the pre-defined planning tasks, the operator can effortlessly retrieve the collected data from the File Management Service. Such an operation can be implemented with ease since a reference path of each generated file path is available in the Mission Component Service.

2.2.3.4 Mission- & asset-centric data storage/indexing and validation service

The Mission and Asset centric Management module is related to the management of data assets that are either prerequisites before the execution of a mission or task (such as the inspection plans, mission and inspection definitions, and the robotic vehicle checklist), as well as inspection generated data (sensor readings, robotic vehicle trajectories, images and video). According to this general concept, it conforms to the I&M-related data sources and therefore facilitates their efficient accessibility through the structured DMS RESTful API services. Additionally, it is responsible for performing the corresponding filtering of the collected observations and association with the respective Mission or asset-specific entities. Finally, it supports an internal validation process, where the incoming information is evaluated according to its data and metadata structure, the respective data formats, and its valid attribution to the corresponding entity. The following bullet points will attempt to briefly provide descriptions of the operations that are executed by using this service.

- *Registration of an Inspection Plan*: By the definition of a new inspection plan along with the corresponding tasks, the user retrieves from the mission component service details related to the asset, robotic system and payload, as well as the personnel so as to formulate the new plan. When the details are filled in and submitted, the new inspection plan is validated against predefined validation rules and is subsequently stored in the Mission and Asset-centric Storage Service.

- *Retrieval of an Inspection task definition*: After filtering and selecting the target asset via the web data portal, the operator selects the available inspection tasks associated with this asset. Selecting the target task definition, all the associated details are fetched from the Mission and Asset-centric Storage Service.
- *The on-robot camera captured image storage in the DMS*: After the successful execution of a mission and mission task, the images captured by the robotic systems are stored in the DMS. The image metadata is uploaded to the Mission/Asset centric storage Service and associated with the actual image file, which is stored in the File Management Service.

2.2.3.5 AI metadata service

The AI metadata service is responsible for including and storing the original inspection images, along with the results that occurred after the execution of the AI defect detection models, and their metadata. The outcome of this functionality is the successful relation of the above-mentioned information, where the initial collected data will be updated, in regards to their initial metadata and the metadata will contain the results of the AI model (i.e., the bounding box, the detected defect type, and the confidence level of the model's prediction), denoted as annotation information in the properties of the specific image.

2.2.3.6 Static sensors storage service

The Static sensors service is responsible for storing and indexing the timeseries observations that are retrieved by the IoT wireless sensor networks (WSN) that have been installed in the pre-defined location of the registered infrastructure assets. In specific, wireless accelerometer sensors (i.e. using cellular telecommunication connections Long-Term Evolution (LTE)) were deployed in the viaduct and refinery pilot cases and thus transmitting the timeseries observations under the standardised JavaScript Object Notation (JSON) data format, through the installation of physical gateways and the most-known transfer protocols (i.e., Hypertext Transfer Protocol (HTTP) RESTful APIs or Message Queuing Telemetry Transport (MQTT)). Sequentially, well-established procedures such as the harmonisation and the data transformation have been scheduled in order to successfully store them in the SensorThings API. Various filtering and data query capabilities are offered to extract the data collected over an asset. Some of them are provided below.

- **$filter:** allows clients to filter a collection of entities that are addressed by a request URL.

This process is facilitated through the generation of an authorisation code, which is exchanged in order to receive the access token (JWT). Finally, the DMS Back-End verifies the token, evaluating its format, signature, standard claims, and application permissions, and thus permits the protected exploitation.

To accomplish this workflow, an email and a simple email transfer protocol (SMTP) server were created and Keycloak was configured to send email invitations through them. The identity Manager offers user interfaces for changing its configuration, creating new users, creating new roles and login/logout purposes. Also, relevant interfaces exist for defining certain user management policies and role mappings. For the needs of the PILOTING project, only a single role was necessary to be created (entitled "piloting-admin") for restricting administrative privileges to a few users (i.e., the Administrators).

2.2.5 Implementation and Evaluation

This section provides information on the implementation and hosting of our solution, and the technologies that have been utilised to address the different functionalities based on previously described user requirements. Eventually, results of the experimental evaluation of the above-mentioned functionalities are also provided.

2.2.5.1 Implementation and hosting

The Piloting DMS and Authentication systems have been implemented on top of several open-source tools and libraries, and have been deployed and currently hosted in two separate Ubuntu machines (Ubuntu 20.04.2 LTS). The DMS is hosted in a 64-bit machine with two dual-core Intel CPUs, 8GB of RAM, and 48GB of storage. Whereas, the Authorisation system is hosted in a 64-bit machine with four dual-core Intel CPUs, 8GB of RAM and 20GB of storage. Details of the technical characteristics and the modules that each of the aforementioned systems contain are described in Table 2.2.

Subsequently, the Sensing part of the SensorThings API has been employed, using the open-accessed deployed framework of the FROST Server (FRaunhofer Opensource SensorThings-Server) [28]. The deployment of the SensorThings API was implemented in Docker v20.10.14, installed in an Ubuntu 20.04.2 LTS machine. It consists of two Docker containers, with the first containing the Fraunhofer's Opensource SensorThings API, utilising fraunhoferiosb/frost-server-http:latest image and the second containing a PostGIS database, which is a spatial database extender for PostgreSQL object-relational database, utilising postgis/postgis:11-2.5-alpine image. The root URI of the service is https://piloting-sensorthings.iccs.gr/FROST-Server/v1.1/. It provides as a response a JSON object with a property named value. The value of this property is a JSON Array containing one element for each entity set of the SensorThings Service.

Table 2.2. Specification of the PILOTING DMS operational environment, software and frameworks used with their respective versions.

PILOTING-DMS	PILOTING-Authentication and User Management
• URL: https://isense-piloting.iccs.gr/ • OS: Ubuntu 20.04.2 LTS (GNU/Linux 5.4.0-90-generic x86_64) • SSL/HTTPS: Enabled • Nginx version: nginx/1.18.0 (Ubuntu) • Python version: 3.8.10 • Django version: 3.2.6 • Gunicorn version: 20.1.0 • PostgreSQL version: 12.9	• URL: https://piloting-auth.iccs.gr/ • OS: Ubuntu 20.04.2 LTS (GNU/Linux 5.4.0-94-generic x86_64) • SSL/HTTPS: Enabled • Nginx version: nginx/1.18.0 (Ubuntu) • Openjdk version: 1.8.0_312 • Keycloak version: 7.0.0

2.2.5.2 Performance evaluation

All the experiments related to the communication between the main software modules of the PILOTING were implemented using Postman. Several Postman collections were created with a sequence of requests, which efficiently simulated a full robotic inspection lifecycle with experimental data. Furthermore, we evaluated the continuous availability and uptime of the DMS APIs, by employing an external online service tool, namely StatusCake [29] and performed repetitive HTTP requests for seven consequent days in order to evaluate the performance of the server availability, which reached the 100%. Subsequent evaluation tests were performed to verify that users will be able to upload and download large files in a short time. Hence, the overall uploading time of a Very High Resolution (VHR) image to the DMS API was 12.28 s and 17.57 s the corresponding downloading time.

Besides the aforementioned experimental testing cases, both DMS and SensorThings APIs were tested for their efficiency, security and reliability during the first pilot phase of the project, which lasted approximately one week per case. Table 2.3 showcases the overall results regarding the number of inspection plans, tasks, missions, robotic systems and payloads, and therefore the collected data during pilots. To perform all these interconnections and intercommunications with several software and tools, and thus handle all the generated data sources, inevitably a mandatory step to make is to be granted as an authorised and authenticated user. Thus, for the needs of the project, a non-neglected number of 22 unique user accounts were created and delivered through email invitations, with this number expected to be substantially increased in the long run of the project.

Table 2.3. The overall number of the generated inspection plans, missions, and the collected data during pilot demonstrations.

	Pre-mission phase				During-mission phase		Post-mission phase
	Assets	Inspection Plan	Robotic Systems	Payload	Missions	Mission Tasks	File Metadata
Refinery	5	6	3	9	4	4	8
Viaduct	2	3	2	4	8	22	19
Tunnel	7	6	2	1	4	19	13
TOTAL	**14**	**15**	**7**	**14**	**16**	**45**	**40**

2.2.6 Software and Data Availability Statement

The DMS API course code is available in the official GitLab repository of ICCS, under the open-accessed license of Apache 2.0, (https://isense-gitlab.iccs.gr/piloti ngext/piloting_dms). Subsequently, the detailed documentation of the DMS API end-points and responses, as well as the DMS customised data models along with the full definition of the entities are provided through the Zenodo community of the PILOTING, in the following URLs (DMS API documentation: https://zeno do.org/record/7351723#.ZDA2EnbP1D9 and DMS Customised Data model: https://zenodo.org/record/7351760#.ZC_x2HbP1D9).

2.3 Data Distribution and Harmonization Layer (DDHL)

2.3.1 Introduction

In the realm of modern technology and data-driven systems, effective data management and distribution play pivotal roles in facilitating seamless communication between diverse components. The Data Distribution and Harmonization Layer (DDHL) emerges as a central and indispensable entity within the PILOTING I&M platform. This document embarks on a comprehensive exploration of the DDHL, delving into its architecture, functionalities, and significance in bridging the communication gaps that often arise in complex system landscapes.

The DDHL serves as a linchpin in the PILOTING I&M ecosystem, orchestrating the collection, harmonization, and dissemination of data across its various modules. By acting as an intermediary, the DDHL facilitates the seamless exchange of critical information among interconnected systems. At its core, the DDHL is tasked with providing vital information to the General Robotic Control Stations (gRCS) for an Inspection Plan to be performed, collecting both generic data from the gRCS and mission-specific data from individual Robotic Control Systems (RCS). This amalgamated data is subsequently uploaded to the Data Management System

(DMS), a repository accessible to both the Intelligence and Visualisation portal (IVP) and AI services.

An essential capability of the DDHL is its ability to transform heterogeneous data formats, inherent in different robotic systems, into a standardized structure. This transformation enables cohesive integration of data from disparate sources, eliminating compatibility obstacles that often hinder effective collaboration. Beyond data collection and distribution, the DDHL's role extends to the culmination of missions, where it facilitates the efficient flow of information between AI Systems and the platform. By utilizing RabbitMQ, the DDHL triggers notifications to the AI Systems, conveying comprehensive information about captured images that are earmarked for AI analysis. The results generated by AI Systems subsequently flow back to the DDHL, initiating a process in which image metadata is enriched through a dedicated Conversion Tool, thereby enhancing the annotations associated with harmonized images. Moreover, the DDHL serves as a conduit for the incorporation of data from external Internet of Things (IoT) sensors. Leveraging distinct methods, it acquires, transforms, and stores IoT-generated data within the SensorThings API, ensuring a cohesive data landscape.

This document explores DDHL's technical underpinnings, including its containerized services managed by an NGINX server, which operates as a proxy and load balancer. The DDHL employs RESTful APIs to foster seamless communication between various elements, such as the gRCS, RCS, IoT Sensors, DMS, AI Systems, and SensorThingsAPI. The development of the DDHL leverages the Python programming language and the Django REST framework, both of which contribute to its reliability and extensibility. Consistent with the overarching spirit of the PILOTING project, the DDHL's source code is slated for public release through INLECOM's GitLab repository, adhering to the open-access ethos by adopting the MIT license. Practical deployment strategies are addressed in this document, offering insights into utilizing containerization through a docker compose file. This configuration encapsulates essential services, including RabbitMQ and a pre-configured DDHL image that incorporates the Conversion Tool. Guidance is also provided on configuring DMS-specific variables, credentials, and RabbitMQ parameters via environment files.

2.3.2 Development Framework of the I&M PILOTING Data Distribution and Harmonization Layer

2.3.2.1 PILOTING workflow

The elucidation of a platform's workflow assumes paramount importance in the comprehensive understanding of the functionalities inherent to the data distribution and harmonization component (Figure 2.1). The workflow configuration serves as the foundational structure that governs the movement and orchestration of data within the platform. It is the essential canvas upon which the intricate details of data distribution are painted. Without a meticulous comprehension of this workflow, one would find it exceedingly challenging to discern the role and significance of the data distribution component within the broader system. The workflow elucidates the systematic progression of data, delineating its stages from acquisition and transformation to storage and dissemination.

In this section, a comprehensive analysis of the PILOTING project's workflow unfolds, delineating three distinct and pivotal phases: pre-mission, during mission, and post mission. Each of these phases comprises a critical segment in the project's overarching narrative, engendering unique insights and challenges inherent to the journey of piloting.

Pre-mission phase

The first step is to build a digital representation of the desired asset.

1. **Establishing assets:** In the realm of asset management, the initial step involves the creation of assets. This encompasses two distinct scenarios, as mutually agreed upon:
   a. In cases of Pressure vessels, Pipes, and Storage Tanks, the asset model is shaped within the Intelligence and Visualization Portal (IVP) utilizing the AssetBuilder tool. The model is then exportable in a pre-determined format (STL).
   b. For scenarios involving refinery Ground-monitoring, Tunnels, and Viaducts, a three-dimensional asset model is crafted during an initial reconnaissance mission. This mission results in the generation of a point cloud file, adhering to an agreed format (PCD). Subsequently, this file can be uploaded through the IVP, and further processed to facilitate viewing through web interfaces.
2. **Identifying inspection locations:** Upon the availability of the asset model, engineers are empowered to annotate it using a specialized tool embedded within the IVP. An inspection location, serving as an annotation on the model, can be defined in three distinct ways:
   a. Inspection Points: Selectable and maneuverable points, visually depicted as small orbs or spheres. These points are initially positioned on the surface of the asset structure.
   b. Inspection Areas: Formed by outlining bounding boxes that encompass specific sections of the asset's surface. In mesh models, the area is defined as the intersection between the bounding box and the model's geometry.

For point cloud models, the area is delineated by points enclosed within the bounding box.

c. Inspection Parts: In the case of AssetBuilder models, it's feasible to pinpoint specific parts of the asset, such as a Manway on a pressure vessel, as inspection locations.

3. **Attaching Asset-Related Documentation:** Engineers are also granted the ability to attach diverse media files associated with the asset. These encompass files like asset drawings, inspection manuals, technical specifications, inspection safety plans, and others.

At this juncture, all pertinent asset information is not only established within the IVP but is also preserved within the DMS. Subsequently, the focus shifts towards the formulation of an inspection plan.

4. **Crafting an inspection strategy:** The engineer embarks on crafting an inspection plan for the asset. This plan outlines the key aspects of the inspection, including its scope, location, objectives, methodology, and more. Additionally, the plan integrates both physical and digital inspection equipment required for the mission, encompassing robotic systems, sensor payloads, and other necessary tools. Furthermore, it incorporates a comprehensive suite of inspection-related documents, encompassing safety protocols, regulatory compliance, exclusion zones, standard and emergency procedures, risk assessment reports, and all correlated inspection tasks. Each individual plan may encompass an array of distinct inspection tasks, each intricately tied to specific inspection locations embedded within the infrastructure assets.
5. **Defining inspection tasks:** The engineer proceeds to outline and define the tasks that need to be executed at precise locations on the asset (Figure 2.6). It's conceivable to establish multiple inspection tasks at a given inspection location, carefully sequencing them to facilitate the Robotic Control System (RCS) in constructing a meticulously planned trajectory prior to the mission.
6. **Incorporating inspection files:** The engineer is also empowered to incorporate various files that hold relevance to the ongoing inspection plan. Examples include risk assessments, safety guidelines, emergency procedures, and exclusion zone details.
7. **Assigning personnel:** A crucial phase involves the engineer assigning the roster of personnel required for each individual inspection task.
8. **Designating robotic systems:** In this phase, the engineer allocates specific robotic systems for deployment during each inspection task. Notably, the DMS holds a comprehensive repository of all available robotic systems and their corresponding payload information. Within the IVP, engineers simply select the most suitable equipment from this repository.

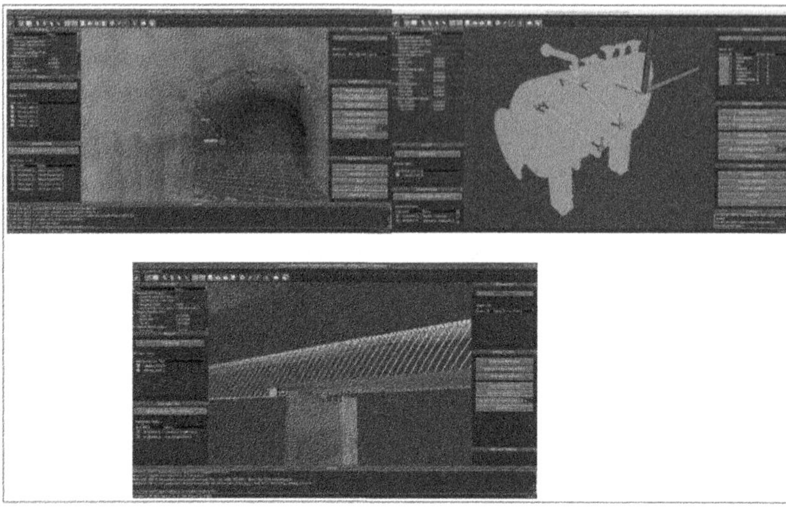

Figure 2.6. Integration of gRCS with multiple robotic vehicles during inspections of a tunnel (top left), vessel (top right) and viaduct (bottom).

This phase incorporates the earliest stages of the inspection planning process, in which 3D models, inspection locations and complementary files are imported/created in the visualisation interface and are available in the DMS. With these data sources available, entities related to the inspection plan, inspection tasks, files, the assigned personnel, robotic system and payloads can be defined along with information about inspection locations and orders of mission tasks' execution.

During-mission phase

1. Preceding the execution of a mission, the Robotic Control System (RCS) is mandated to establish connectivity with each robotic system engaged within, see Figure 2.6, the given use case and conduct a rigorous assessment of said connectivity. Furthermore, both the Ground Robotic Control System (gRCS) and the Robotic Control System (RCS) shall undergo a meticulous evaluation of their respective connections to the Data Distribution and Harmonization Layer (DDHL).
2. In order to instigate a mission, the gRCS must formally solicit the retrieval of inspection plan data from the DDHL. This solicitation necessitates the provision of a unique inspection plan identifier, which serves as the exclusive identifier for the intended inspection plan. Significantly, the operator of the gRCS obtains all the Inspection Plans related to a particular site category. This process empowers the gRCS to select the Inspection Plan it requires, enabling it to access crucial information such as the Unique Identifier of the Plan and its associated tasks.

3. Subsequent to the mission initiation request, the DDHL assumes the responsibility of forwarding this request to the Data Management System (DMS), subsequently routing the response from the DMS back to the gRCS.
4. The DMS, upon receipt of a request for an inspection plan from the DDHL, is programmed to expose all requisite information in JSON format via Rest-Full APIs. The JSON data, inclusive of the complete pathway to the DMS from which the gRCS/mRCS can retrieve each file, encompasses a comprehensive spectrum of data, encompassing Inspection Tasks, Inspection Types, Inspection Locations, Assets, and ancillary metadata or files associated with these entities.
5. The gRCS is tasked with the capacity to discern mission-specific tasks for each mission, facilitate the transfer of necessary mission data to the robotic vehicle, and meticulously complete a checklist encompassing inspection tasks that mandate fulfilment prior to mission commencement. This checklist incorporates a suite of verifications conducted in the moments leading up to mission initiation, encompassing assessments such as battery status and sensor functionality. It is crucial to note that the checklist is dynamically sourced from the respective robotic system, tailoring its content to the specific requirements of each system.
6. Throughout the execution of a mission, the gRCS is entrusted with the responsibility of closely monitoring mission progress. In cases necessitating real-time adjustments to the mission plan, the gRCS shall possess the requisite functionality to implement such alterations seamlessly. Furthermore, each RCS assumes a pivotal role in recording and archiving invaluable data originating from on-board sensors, thereby enriching the data repository in the context of the inspection mission.

Post-mission phase

1. Following the completion of the mission, the gRCS will compile all mission-related data results and transmit the entirety of data logs to the DDHL. This dataset encompasses the planned trajectory, the path traversed, the mission checklist, and the event log.
2. Upon receipt of this request, the DDHL assumes the responsibility of creating a mission record and subsequently proceeds to process, enhance, and forward the received data to the Data Management System (DMS), where it will find its permanent storage. As part of this process, each dataset is assigned a synchronized identification (UUID), ensuring the DDHL's ability to correlate and associate gRCS and mRCS data pertinent to the same mission.
3. Additionally, each RCS involved in the mission is required to upload all mission-specific data logs, which encompass photos, videos, and sensor data,

to the DDHL. Leveraging the previously established "sync ID," the DDHL retrieves the data, generates mission task IDs, and augments the mission data before transmitting it to the DMS.

4. At this stage, concurrently or in parallel, the DDHL undertakes the retrieval of data originating from external Internet of Things (IoT) sensors associated with each asset.
5. The raw inspection data, including measurements such as ultrasonic testing (UT), undergoes a harmonization process, ensuring consistency, and is transformed into data formats compatible with the predefined data model of the platform.
6. Data synchronization, facilitated by the DDHL, is conducted at this juncture to harmonize all retrieved data, comprising data from the RCS, robotic systems, and external static sensors. Subsequently, all data is amalgamated into a final data format ready for upload to the DMS.
7. The DDHL, in the ensuing step, uploads all mission-specific data to the DMS.
8. The DMS serves as the repository for all information accumulated during inspection missions conducted by the robotic systems. It further provides access to external services, including AI services and the Integrated Visualization Platform (IVP), making the data available for additional analysis and utilization

2.3.3 Data Distribution & Harmonization Layer Development

The Data Distribution and Harmonization Layer (DDHL) comprises a collection of services that have undergone containerization through Docker and are overseen by an NGINX server, serving the dual purpose of a proxy and a load balancer. Within its framework, the DDHL offers a range of RESTful APIs to facilitate communication and data exchange among various entities, including the Ground Robotic Control System (gRCS), Robotic Control System (RCS), Internet of Things (IoT) Sensors, Data Management System (DMS), Artificial Intelligence (AI) Systems, and SensorThingsAPI, as delineated in Table 2.4. The development of these services is underpinned by the utilization of the Python Programming Language and the Django REST framework.

Adhering to the vision of the PILOTING project the DDHL course code has been published through the official GitLab repository of INLECOM, under the open-accessed license of MIT in due time, before the end of PILOTING project. The Gitlab repo contains the source files of the DDHL component, its prerequisites in the requirements.txt file, the MIT license and a README.md file with technical information. For a local deployment of the DDHL component, the suggested

Table 2.4. DDHL RESTful APIs.

Type	DDHL RESTful APIs	Functionality
GET	apis/Inspectionplan/<site_id>/	Retrieve all Inspection Plans filtered by site (Viaduct, Refinery or Tunnel)
GET	apis/Inspectionplan/<plan_id>/All/	Retrieve and return an Inspection Plan
GET	apis/File/Download/<file_id>/	Retrieve and download a file from the DMS
POST	apis/Mission/gRCS/	Post generic data to the DMS
POST	apis/Mission/mRCS/AEROCAM/	Post mission specific data to the DMS gathered from AEROCAM RCS
POST	apis/Mission/mRCS/TTDRONE/	Post mission specific data to the DMS gathered from TTDRONE RCS
POST	apis/Mission/mRCS/VIAD_DRONE/	Post mission specific data to the DMS gathered from VIAD DRONE RCS (Bearing Inspection)
POST	apis/Mission/mRCS/VIAD_DRONE/Installation/	Post mission specific data to the DMS gathered from VIAD DRONE RCS (Viaduct surveying target or sensor box installation)
POST	apis/Mission/mRCS/ETH/	Post mission specific data to the DMS gathered from UGV(ETHZ) RCS
POST	apis/Mission/mRCS/BIKE/	Post mission specific data to the DMS gathered from BIKE RCS
POST	apis/Mission/mRCS/SATELITE/	Post mission specific data to the DMS gathered from SATELITE RCS
POST	apis/Mission/mRCS/CART/	Post mission specific data to the DMS gathered from CART RCS
POST	apis/Mission/mRCS/AEROX/	Post mission specific data to the DMS gathered from AEROX RCS
POST	apis/Mission/mRCS/RISING/	Post mission specific data to the DMS gathered from RISING RCS
PATCH	apis/Mission/AI/ImageMetadata/	Patch image metadata with AI results

procedure is through containers. Specifically, a docker compose file is used. The file contains the RabbitMQ service which is required by the DDHL and uses a pre-built image of the DDHL that includes the Conversion Tool dependency. Additionally, an environment file must be provided to define the DMS specific variables, credentials and RabbitMQ parameters, as described in README.md file of the repository. To pull and run the compose stack follow the instructions of the code block below.

```
$ git pull <ddhl-repo>
$ docker-compose –env-file .env.local pull
$ docker-compose up -d
```

Access the API at http://localhost:3040/

2.3.4 Data Distribution & Harmonization Layer Architecture

This section delves into the interaction mechanisms of the Data Distribution and Harmonization Layer (DDHL) with both the intrinsic constituents of the I&M PILOTING platform (namely, the Ground Robotic Control System (gRCS) and the Data Management System (DMS)), alongside several extraneous entities, including Robotic Control Systems, Internet of Things (IoT) Sensors, Artificial Intelligence (AI) Systems, and SensorThings API, as show in Figure 2.7.

2.3.4.1 Integration with the internal components of I&M PILOTING platform

This section encompasses the discourse on the Data Distribution and Harmonization Layer's (DDHL) interaction with the internal components of the I&M PILOTING platform, specifically, the Ground Robotic Control System (gRCS)

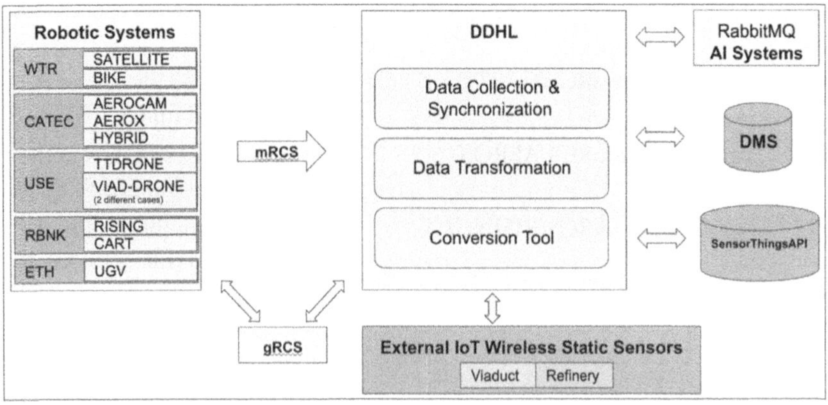

Figure 2.7. Architecture Diagram describing the interconnection of the DDHL with internal components of I&M PILOTING platform as well as external components.

and the Data Management System (DMS). During the pre-Mission phase, meticulously examined, the gRCS initiates communication with the DDHL to retrieve all Inspection Plans by type of the Inspection Site, fetch a specific Inspection Plan, download Point Cloud file, instate Mission and Mission Tasks entities, and transmit generic data through the employment of the DMS's APIs. In the event of an unforeseen setback, the DDHL is primed to effectuate a rollback of the database alterations and subsequently apprise the gRCS of the encountered failure. For a more comprehensive understanding, additional details can be found within the subsequent figures.

Table 2.5. Integration of DDHL with gRCS and DMS.

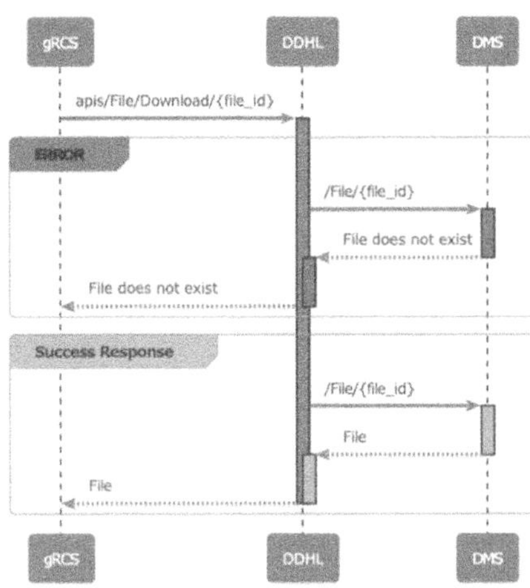

gRCS downloads a file from the DDHL.

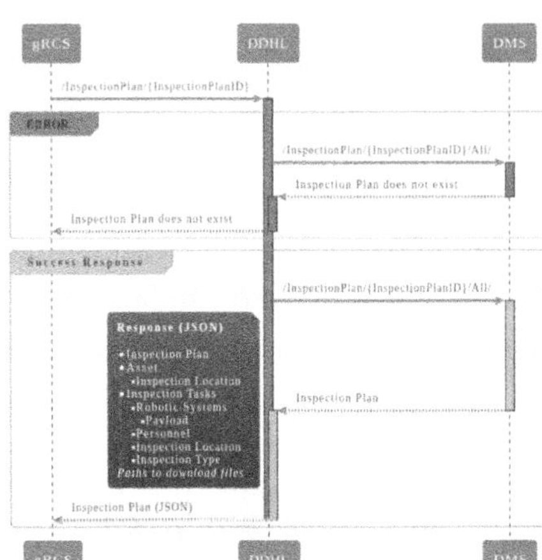

gRCS retrieves an Inspection Plan by providing the Inspection Plan's Unique Identifier to the DDHL.

Table 2.5. Continued

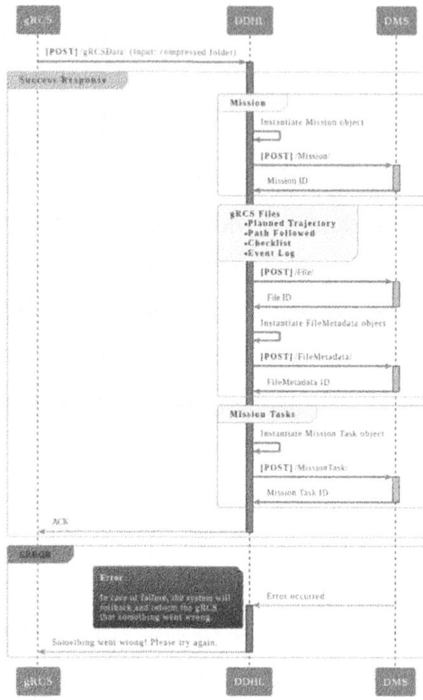

gRCS posts generic data of the Inspection to the DDHL. The DDHL instantiates the Mission and Mission Tasks and posts data to the DMS.

2.3.4.2 Integration with the external components

2.3.4.2.1 Integration with robotic control systems

As previously elucidated, there exists a distinction between two categories of data—generic and mission-specific—each essential in facilitating the execution of a mission. The generic data serves the purpose of initializing a mission during the pre-Mission phase, while mission-specific data encompasses the information collected during the mission's execution. Beyond the standardized interface represented by the Generic Robot Control Station (gRCS), the Robot Control Station incorporates specialized interfaces, denoted as Mission Specific (mRCS), tailored to the particularities of the on-board sensors of the respective robotic vehicle. Consequently, this section delves into the examination of the interplay between the Robotic Systems and the Data Distribution and Harmonization Layer (DDHL). To enable the upload of mission-specific data from each of the Robotic Systems, the DDHL provides dedicated RESTful APIs for each robot (as detailed in Table 2.5). These APIs are designed to receive compressed files containing mission-specific data as input, facilitating the seamless exchange of crucial information.

The Data Distribution and Harmonization Layer (DDHL) assumes responsibility for a multifaceted sequence of actions, commencing with the unzipping of a folder, followed by data transformation and harmonization to align it with predefined formats. Subsequently, the DDHL orchestrates the upload of this data into the appropriate mission context, utilizing the Application Programming Interfaces (APIs) of the Data Management System (DMS). Upon completion, the DDHL furnishes an acknowledgment denoting either success or failure.

In the course of mission instantiation, a Synchronization ID (SyncID) is generated and assigned to the mission. During the mission-specific data upload phase, the DDHL employs this SyncID to synchronize the data seamlessly. Specifically, the DDHL searches for a mission linked to the Inspection Plan provided by the Robotic Control Station (RCS), which in turn corresponds to the SyncID. In cases of unexpected failure, the DDHL initiates a rollback mechanism to reverse the alterations made to the database and subsequently communicates the failure to the respective Robotic System.

For all RCS scenarios delineated herein, the DDHL executes a set of common tasks independent of the specific robotic case. Primarily, it begins by validating the format of the configuration file, with a failure leading to a validation error. Subsequently, the DDHL retrieves the mission associated with the provided Inspection Plan linked to the provided SyncID, along with its corresponding Mission Tasks. It then verifies that all Inspection Tasks listed in the configuration file are congruent with the Inspection Plan provided. Any discrepancy between an Inspection Task and the Inspection Plan generates an error, signalling to the RCS that the mission should exclusively execute tasks aligned with the Inspection Plan.

Continuing its process, the DDHL proceeds to transform and upload mission-specific data, associating it with the Mission and Mission Objects created within the DMS during the upload of generic data. The treatment of each data file type varies concerning harmonization. In a general context, the DDHL generates a new record within the File Table of the DMS for each media file (e.g., image, point cloud, UT Scan measurement) and an additional record in the MetadataFile table for its corresponding metadata. For metadata files supplied by the Robotic Systems, containing data correlated with unique Inspection Task IDs, the DDHL validates the linkage of the provided Inspection Task with the implemented Inspection Plan. It then identifies the associated Mission Task relative to the given Inspection Task. Subsequently, the metadata file, containing the file location within the DMS, is linked to the respective Mission Task in accordance with the specific case. Importantly, in terms of transformation, it is pertinent to note that the DDHL accepts various date and time input values but posts solely in UTC timestamps format (e.g., 2022-11-22T07:13:30.992827Z).

The final phase entails enhancing the Mission Object within the DMS, enriching it with mission-specific information such as the execution datetime and details regarding the sensors involved in mission implementation. Concurrently, the mission is marked as completed.

2.3.4.2.2 Integration with AI systems

The primary duty of the Data Distribution and Harmonization Layer (DDHL) resides in notifying the AI systems upon the completion of a mission task, along with the provision of comprehensive data required for retrieving mission task images. Subsequently, the AI services undertake the crucial process of identifying defects within the scrutinized regions of each image.

Moreover, the DDHL is obligated to collect and synchronize annotations generated by the AI systems, enhancing the metadata associated with each image stored within the Data Management System (DMS) with these annotations.

The interaction between the DDHL and AI systems transpires via a message broker mechanism, specifically RabbitMQ, a widely adopted open-source message broker. To cater to the distinct PILOTING use cases encompassing Refinery, Viaduct, and Tunnel scenarios, three distinct queues have been established, as illustrated in Figure 2.8. In the event where a Robotic System submits mission-specific data to the DDHL, the subsequent course of action is depicted in Figure 2.9. Upon the successful completion of the data upload, the DDHL will dispatch a notification to one of three designated queues, contingent upon the specific inspection site in question. The aforementioned message shares information on the images captured during the mission, such as the location of the images on the DMS and their metadata.

This process serves as a notification mechanism for the AI System, which has subscribed to these queues, signifying the culmination of a mission. Subsequently, the AI System proceeds to directly retrieve all mission images from the Data Management System (DMS) and initiates the execution of the requisite algorithms. Following the analysis, the AI System transmits the outcomes in the form of annotations to the DDHL through a RESTful API. This API mechanism is designed to enable each AI system to enrich image metadata with the annotations it has generated.

In the concluding stage, the DDHL proceeds to relay the annotations to the Conversion Tool, a software component tasked with the standardization of output generated by all AI services employed across the various PILOTING use cases. Then, the DDHL updates the metadata objects of the images on the DMS with the provided annotations.

Figure 2.8. Available queues on the RabbitMQ for PILOTING use cases.

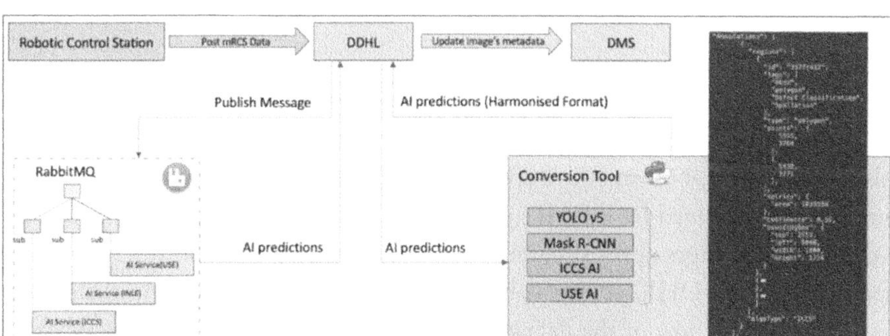

Figure 2.9. Architecture Diagram describing AI System triggering, Harmonisation of AI service's output and updating image's metadata on the DMS.

2.3.4.2.3 Integration with IoT wireless static sensors network & SensorThings API

In this chapter, we delve into the intricate process of data communication between the Data Distribution and Harmonization Layer (DDHL) and Static IoT Sensors, with a primary focus on the utilization of the SensorThingsAPI standard. The empirical evidence is drawn from two compelling pilot cases, Refinery and Viaduct, which serve as exemplars, illustrating the robustness of this approach in real-world scenarios.

2.3.4.3 General framework

The seamless integration of IoT sensors and the DDHL system constitutes a fundamental pillar of contemporary infrastructure management and data-driven decision-making. Within this section, we dissect the underlying communication framework and data integration methodology.

DDHL features a module operating at regular intervals, responsible for the collection of data from designated IoT sensors. Notably, this framework does away with conventional mission-centric approaches and instead associates sensors with assets, such as refineries or viaducts within the infrastructures. This asset-centric approach ensures the continuous acquisition of sensor measurements, independent of mission durations, ultimately storing them as observations of a DataStream.

To ensure alignment with industry standards and facilitate seamless integration, data collected by DDHL undergoes a transformation process, conforming to the SensorThingsAPI standard. As elucidated below this standardized format encapsulates key attributes, including an identifier for the relevant DataStream, measurement values, and the timestamp of acquisition.

2.3.4.3.1 Refinery case

Within the Refinery pilot case, accelerometer sensors are strategically deployed both indoors and outdoors. The transmission of measurement data to DDHL is orchestrated through a broker, employing the MQTT protocol. This protocol is used to publish data in JSON format to a specific topic. Both the gateway and the backend components, managed by DDHL, embrace this protocol. The gateway subscribes to dedicated messages approximately every five minutes from the aforementioned topic, while the DDHL, acting as backend, is responsible for storing the sensors' observations. The flow of the data acquisition for the Refinery IoT Sensors case is depicted in Figure 2.10.

Post data acquisition, DDHL assumes the responsibility of data consumption and conversion into the standard format of an Observation entity of the SensorThingsAPI. This standardized format comprises the DataStream identifier associated with the corresponding sensor in the database, the set of measurement

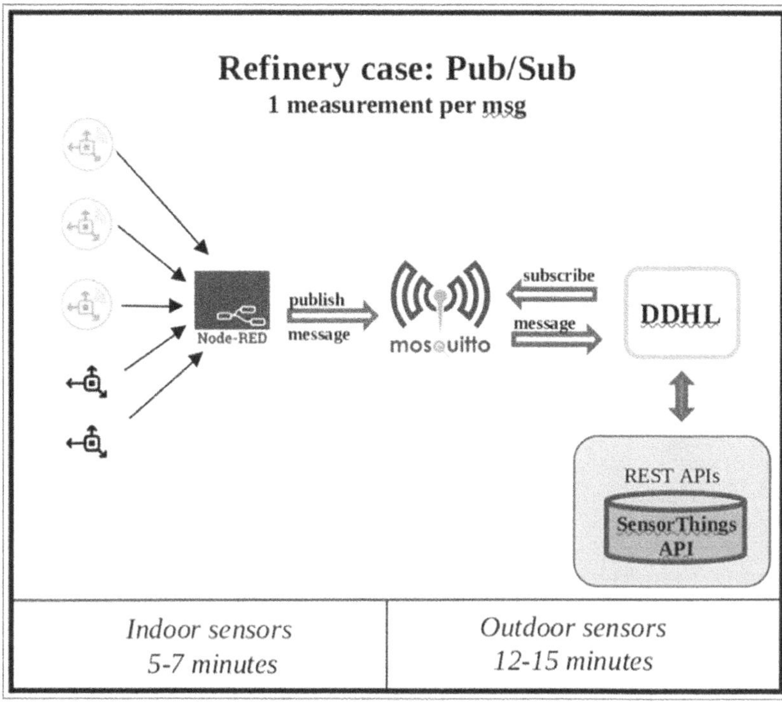

Figure 2.10. Flow of the data acquisition for the Refinery IoT Sensors case.

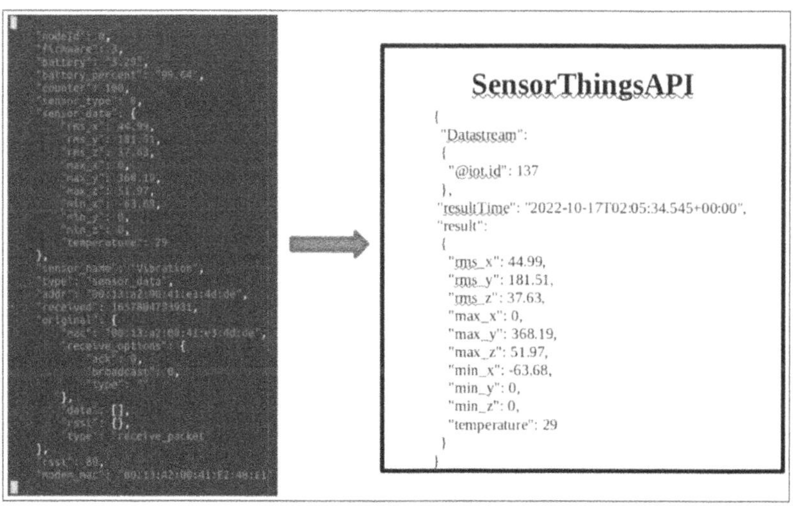

Figure 2.11. Example of measurement data transformation in the Refinery case.

values, and the timestamp of measurement acquisition. These observations are then posted, through an API call, to the SensorThings database, where they are automatically linked to the correct DataStream and corresponding sensor. An example of a measurement data transformation is depicted in Figure 2.11.

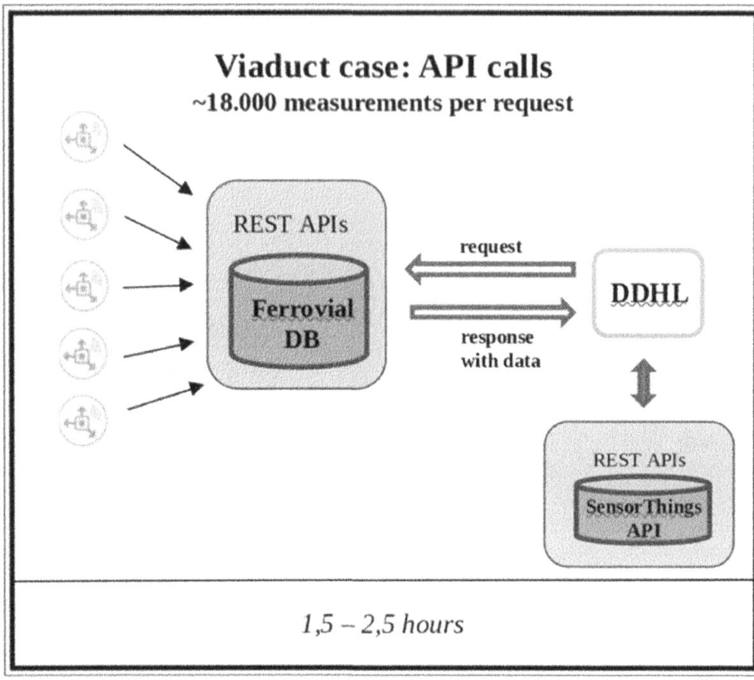

Figure 2.12. Flow of the data acquisition for the Viaduct IoT Sensors.

2.3.4.3.2 Viaduct case

In the Viaduct pilot case, five sensors of a single type, namely accelerometers, have been installed on site. Upon a successful connection, the DDHL delivers the collected data from the accelerometer sensors to the backend, by utilizing a RESTful API that is provided by the platform to facilitate the data extraction. Each sensor within the Viaduct environment generates an extensive volume of data – approximately eighteen thousand measurements over a brief three-minute interval. The flow of the data acquisition for the Viaduct IoT Sensors case is depicted in Figure 2.12.

To ensure data accuracy and eliminate redundancy, DDHL employs a timestamp-based mechanism that compares the timestamp that designates the start of each interval with the most recently available timestamp in SensorThingsAPI for each sensor, guaranteeing the posting of only novel data.

For each relevant datetime, DDHL retrieves the corresponding data, crafting observations that rigorously adhere to the SensorThingsAPI standard. These observations are then promptly posted to the SensorThings database, thereby guaranteeing their linkage with the accurate Datastream and corresponding sensor. An example of a measurement data transformation is depicted in Figure 2.13.

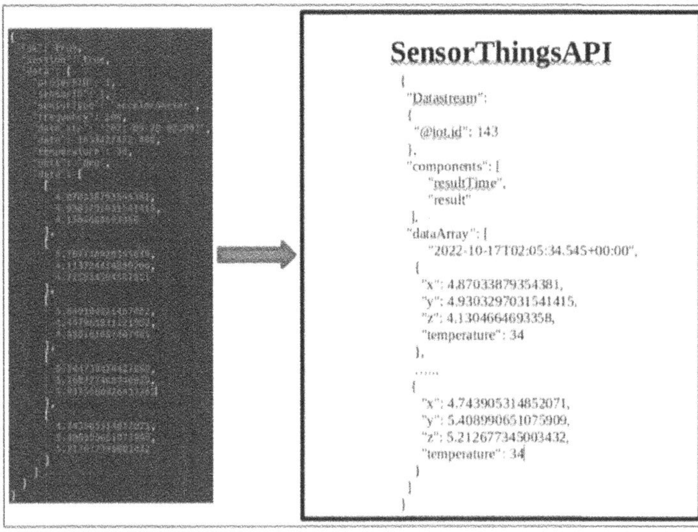

Figure 2.13. Example of measurement data transformation in the Viaduct case.

2.3.5 Software and Data Availability Statement

The DDHL API course code is available in the official GitLab repository of INLECOM, under the open-accessed license of MIT, (https://gitlab.com/piloting-h2020/data-distribution-and-harmonization-layer/), as well as all the detailed documentation, models and all relevant information of the DDHL.

2.4 The General Robot Control Station

Having made the first general introduction to the operation of the I&M Platform, we will now go on to describe the gRCS in more detail. In the following lines we will describe its software architecture with emphasis on the software modules and the communication between them (Figure 2.14), the graphic and finally, we will analyse the communication with the robotic platforms.

2.4.1 Software Modules

In this section all the sub-modules that are part of the gRCS software architecture will be identified and described. Additionally, it will be described how communications are carried out between the different modules of the system (Figure 2.14).

2.4.1.1 Data model

This foundational module manages the storage of essential information generated during gRCS operation. It employs Data Transfer Objects (DTOs) to store and

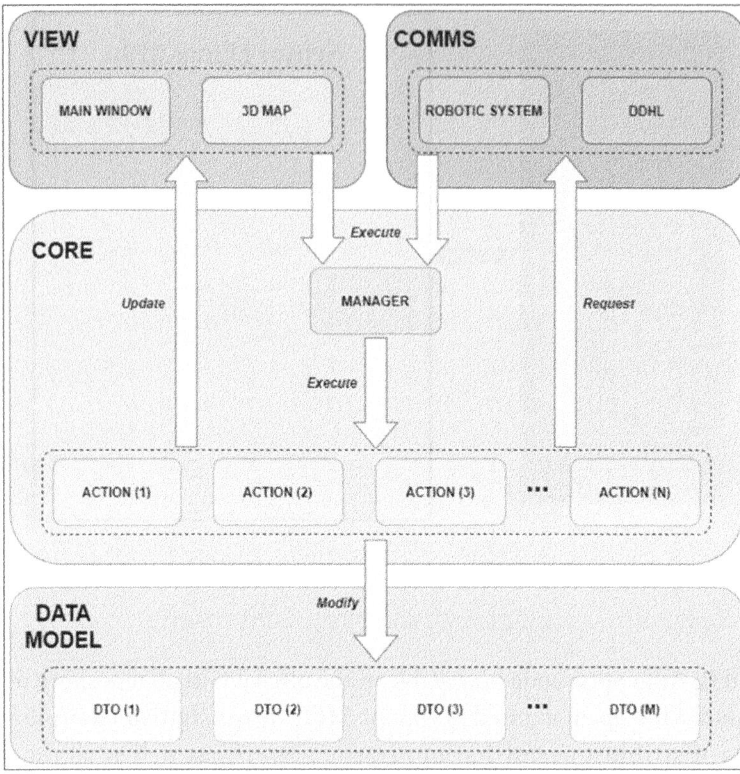

Figure 2.14. Software architecture.

transmit data between system modules. These DTOs inherit from a common interface, simplifying communication between modules.

2.4.1.2 Core

Located at the center of the architecture, this paramount module makes decisions and coordinates actions across the system. It comprises sub-modules called ACTIONS, responsible for executing specific tasks like modifying a planned path. An ACTION involves two phases: performing the action and informing higher layers about updates. The MANAGER module chooses and executes ACTIONS based on incoming requests.

2.4.1.3 Communications

The top module handles external communications and enforces communication protocols. Divided into two modules, it manages communication with both the gRCS and robotic systems and the Data Distribution and Harmonization Layer. It also interfaces with lower layers for commanding external subsystem requests and updating the data model.

Communication between gRCS and DDHL is established through RESTful interfaces, ensuring uninterrupted interaction. A synchronization identifier ('sync id') is generated to harmonize overall post-mission data from gRCS with the specific post-mission data from the robotic system. Configuration of DDHL can be easily executed using either a URL or an IP address along with the corresponding port number. Moreover, a newly introduced endpoint simplifies the retrieval of inspection plans categorized by site type, such as Viaduct, Refinery, or Tunnel. Furthermore, the data exchange mechanism between gRCS and DDHL has been enhanced by introducing new endpoints for soliciting inspection plans and uploading generic post-mission data.

2.4.1.4 View

This uppermost module presents a visualization and control interface for gRCS users. It offers mission status visualization through various elements and controls, enabling user interaction. The layer comprises a main window and a 3D map, which allow users to control ongoing missions. It interfaces with lower layers to update the data model with user inputs and to receive updates on data model changes.

In summary, the gRCS system features a modular architecture that facilitates communication, decision-making, external interaction, and user interface for effective mission management.

Next it will be described the communication between the different modules, a summary of the designed user interface, and the communication with the robotics systems.

2.4.2 Communications Between Modules

As previously discussed, the data model consists of simplified objects called DTOs, streamlining communication interfaces between various system modules. The signals and slots paradigm are employed to implement this interface, utilizing the distinctive features of the Qt framework, upon which the gRCS system is built.

In this paradigm, a signal serves as a notification emitted by a module, potentially of interest to other modules, while a slot is a function within a module that responds to a specific signal from another module. This paradigm boasts several traits as seen below (Figure 2.15):

- Signals can carry parameters.
- Slots can receive parameters.
- A single signal can connect to multiple slots.
- Multiple signals can connect to a single slot.

By considering these characteristics and employing the defined DTOs from the data model, the communication interface between gRCS system modules is

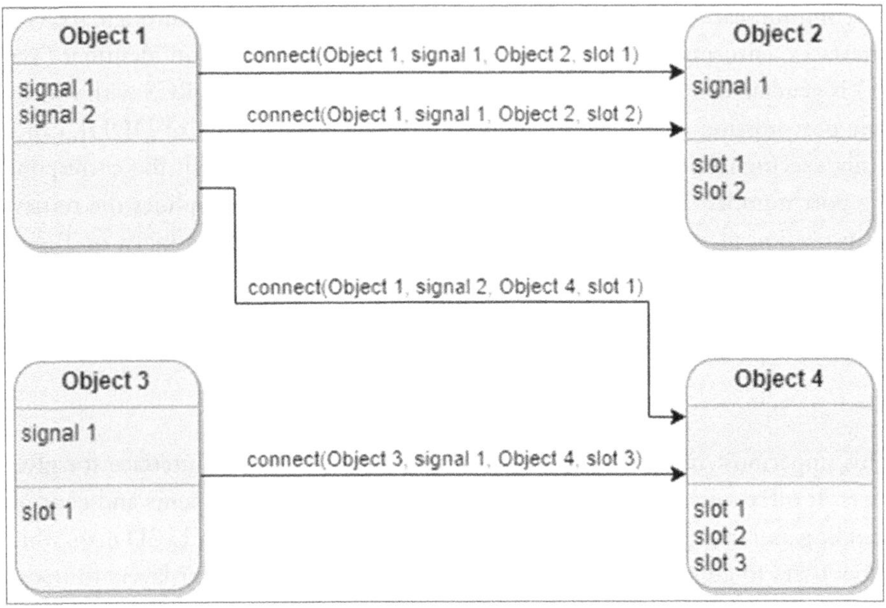

Figure 2.15. Signals and slots paradigm.

simplified, necessitating only a minimal number of signals and slots to be defined. Notably, a connection between a signal and a specific slot is established during the system's boot process. Both the VIEW and COMMUNICATIONS modules possess defined signals that emit when they intend to execute particular actions within the system. These signals link to a slot within the CORE module, responsible for executing these actions.

Furthermore, the CORE module features a signal emitted each time it seeks to update the system about changes in the data model. This signal connects to a slot in the VIEW module, which updates its components accordingly.

Lastly, the CORE module includes a signal emitted when it requires communication-related requests from the COMMUNICATIONS module. This signal connects to a slot within the COMMUNICATIONS module, responsible for managing these incoming requests.

2.4.3 Design of Graphical User Interface

This section explains the graphical user interface (GUI) of the gRCS, its functions, and its design considerations. The gRCS GUI serves to monitor robotic system status and information during PILOTING mode. It features panels that display various received information and enables commanding tasks and mission design. The GUI is independent and can be easily replaced with new interfaces.

Effective GUI design principles include ease of use, intuitive elements (such as windows, buttons, and icons), suitable colour schemes, and a focus on a single primary object. These principles were applied to the gRCS GUI to ensure minimal training for inspection operators.

The subsequent sections detail the gRCS functionalities in a typical operational flow:

- Establishing connection and downloading data from the DDHL server.
- Presenting DDHL-obtained data, like 3D assets, exclusion zones, and inspection tasks, in the 3D viewer.
- Selecting inspection tasks through specific widgets and accessing related information.
- Highlighting the 3D viewer upon task selection.
- Defining mission paths through local path planning in the 3D environment, aided by tools like waypoint adjustment.
- Displaying comprehensive robotic system data, including alarms, pose, and speeds, through windows.
- Providing a console for real-time execution updates.

The included Figure 2.16 offers an overview of the gRCS GUI's structure:

2.4.4 Communication Interface with Robotic Systems

Regarding the communication interface, this module outlines the crucial communication interface employed for data exchange between the general robot control

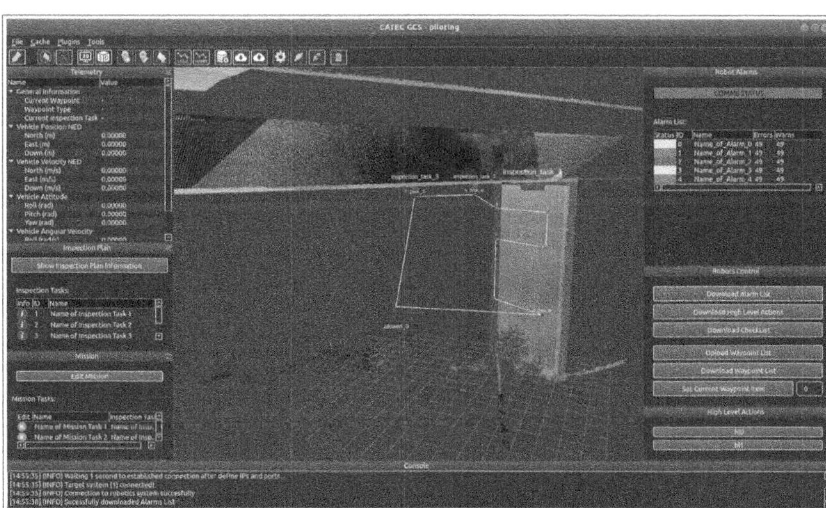

Figure 2.16. General overview of the gRCS graphical interface.

station and the robotic systems. Given that the gRCS oversees real-time mission monitoring, a resilient and consistently open communication connection between the robotic system and the gRCS is imperative. Failures in this interface could lead to the loss of vital mission data during execution.

The communication channel's bandwidth might be limited, necessitating the selection of a communication protocol featuring a lightweight header to maximize data efficiency within each package. Additionally, due to bandwidth constraints, all data sent through this channel must be serialized.

The MAVLink protocol, commonly used in drone-to-ground control station communication, fits these requirements. Designed for low-bandwidth communication and resource-constrained systems, MAVLink ensures data integrity during transmission using CRC-16/MCRF4XX checksum. By leveraging MAVLink's open-source nature, the PILOTING system adapted the protocol to its needs. This involved creating custom messages and sequences while retaining the protocol's fundamental benefits. Custom messages were defined through an XML file, which was then used to generate source code for both the gRCS and robotic systems. Importantly, MAVLink supports multiple programming languages, including C, C++, C#, Java, and Python. This adaptation empowers the communication interface with flexibility and scalability, allowing easy integration of new functionalities. Modifications to the XML file enable the addition of features, with subsequent code generation incorporating these changes. In conclusion, the MAVLink protocol has been chosen as the communication interface due to its inherent advantages and adaptability. By aligning with MAVLink's core benefits and tailoring its messages to suit PILOTING's needs, the system enjoys an effective communication interface without limiting its future growth.

2.4.5 Software and Data Availability Statement

The gRCS course code is available in the official GitHub repository of Fada – CATEC, under the open-accessed license of MIT, (https://github.com/fada-catec/piloting_grcs) as well as all the detailed documentation, models and all relevant information of the gRCS.

2.5 Conclusion

In conclusion, this chapter underscores the significance of meticulous strategic planning, innovative and adaptable architectural design, and strict adherence to universally recognized protocols and standards in the development of cost-effective, proficient, and mature technological solutions for inspection and maintenance frameworks within critical infrastructures. The examination of the PILOTING project's Inspection and Maintenance Platform in this chapter underscores the compelling synergy of integrated technology in the domains of robotics and data management.

Central to this integration, the gRCS fulfils a pivotal role as a critical nexus, facilitating unimpeded communication and coordination among an array of robotic systems, encompassing UAVs, UGVs, and the Inspection and Maintenance Platform itself. Furthermore, the DDHL emerges as a pivotal element in the interconnection between the gRCS and the Data Management System, transforming heterogeneous data into a uniform standardized format, thereby streamlining the seamless flow of information.

Conclusively, the Data Management System assumes the mantle of data custodian, furnishing both safeguarded storage capabilities and expeditious mechanisms for authorized access, retrieval, and systematic indexing of data. This integral component not only ensures data integrity but also establishes an underpinning framework for purposeful analysis and discerning insights within the sphere of inspection and maintenance. In concert, these constituent parts collaboratively establish a resilient ecosystem conducive to efficient and efficacious inspection procedures.

Acknowledgements

This research was funded by the PILOTING project under grant number 871542, supported by the European Union's Horizon 2020 innovation action program.

References

[1] L. Sun, Z. Shang, Y. Xia, S. Bhowmick, and S. Nagarajaiah, "Review of Bridge Structural Health Monitoring Aided by Big Data and Artificial Intelligence: From Condition Assessment to Damage Detection," *J. Struct. Eng.*, vol. 146, no. 5, 2020, doi: 10.1061/(asce)st.1943-541x.0002535.

[2] K. W. C.R. Farrar, "An introduction to structural health monitoring," in *Philos. Trans. R. Soc. London A 365 (1851)*, 2007, pp. 303–315.

[3] L. Gigli, L. Sciullo, F. Montori, A. Marzani, and M. Di Felice, "Blockchain and Web of Things for Structural Health Monitoring Applications: A Proof of Concept," *Proc. - IEEE Consum. Commun. Netw. Conf. CCNC*, pp. 699–702, 2022, doi: 10.1109/CCNC49033.2022.9700679.

[4] S. Khan *et al.*, "Ssvm: An ultra-low-power strain sensing and visualization module for long-term structural health monitoring," *Sensors*, vol. 21, no. 6, pp. 1–16, 2021, doi: 10.3390/s21062211.

[5] V. Kaur, A. Tyagi, M. Kritika, P. Kumari, and S. Salvi, "Crowd-sourcing based android application for structural health monitoring and data analytics of roads using cloud computing," in *2017 International Conference on Innovative Mechanisms for Industry Applications (ICIMIA)*, 2017, pp. 354–358. doi: 10.1109/ICIMIA.2017.7975635.

[6] C. Aguzzi *et al.*, "MODRON: A scalable and interoperable web of things platform for structural health monitoring," *2021 IEEE 18th Annu. Consum. Commun. Netw. Conf. CCNC 2021*, no. Ml, 2021, doi: 10.1109/CCNC49032.2021.9369492.

[7] G. E. Eksi, B. Tekinerdogan, and C. Catal, "Software security management in critical infrastructures: a systematic literature review," *Turkish J. Electr. Eng. Comput. Sci.*, vol. 30, no. 4, pp. 1142–1161, 2022, doi: 10.55730/1300-0632.3841.

[8] Y. Q. Ni, K. C. Lin, L. J. Wu, and Y. W. Wang, "Visualized Spatiotemporal Data Management System for Lifecycle Health Monitoring of Large-Scale Structures," *J. Aerosp. Eng.*, vol. 30, no. 2, pp. 1–14, 2017, doi: 10.1061/(asce)as.1943-5525.0000622.

[9] Q. Liu, G. Lu, J. Huang, and D. Bai, "Development of tunnel intelligent monitoring and early warning system based on micro-service architecture: the case of AnPing tunnel," *Geomatics, Nat. Hazards Risk*, vol. 11, no. 1, pp. 1404–1425, 2020, doi: 10.1080/19475705.2020.1797906.

[10] A. Scullion, "Harnessing open source big data technology to transform structural health monitoring from data logging into decision support information," *9th Eur. Work. Struct. Heal. Monit. EWSHM 2018*, no. July, 2018.

[11] A. Larsson and C. Große, "Data use and data needs in critical infrastructure risk analysis," *J. Risk Res.*, vol. 26, no. 5, pp. 524–546, 2023, doi: 10.1080/13669877.2023.2181858.

[12] C. Große, "Enhanced information management in inter-organisational planning for critical infrastructure protection: Case and framework," *ICISSP 2021 – Proc. 7th Int. Conf. Inf. Syst. Secur. Priv.*, no. Icissp, pp. 319–330, 2021, doi: 10.5220/0010186803190330.

[13] K. Y. Koo, N. De Battista, and J. M. W. Brownjohn, "SHM data management system using MySQL database with MATLAB and web interfaces," in *SHMII-5 2011 – 5th International Conference on Structural Health Monitoring of Intelligent Infrastructure*, 2011, no. December.

[14] L. Alonso, J. Barbarán, J. Chen, M. Díaz, L. Llopis, and B. Rubio, "Middleware and communication technologies for structural health monitoring of critical infrastructures: A survey," *Comput. Stand. Interfaces*, vol. 56, pp. 83–100, 2018, doi: 10.1016/j.csi.2017.09.007.

[15] A. Koubaa *et al.*, "Dronemap Planner: A Service-Oriented Cloud-Based Management System for the Internet-of-Drones," *Ad Hoc Networks*, vol. 86, Sep. 2018, doi: 10.1016/j.adhoc.2018.09.013.

[16] J. A. Rice, K. A. Mechitov, B. F. Spencer, and G. A. Agha, "A Service-Oriented Architecture for Structural Health Monitoring using Smart Sensors," *Proc. 14th World Conf. Earthq. Eng.*, 2008, [Online]. Available: http://osl.cs.illinois.edu/docs/wcee08/wcee08.pdf.

[17] J. Giao, A. A. Nazarenko, F. Luis-Ferreira, D. Gonçalves, and J. Sarraipa, "A Framework for Service-Oriented Architecture (SOA)-Based IoT Application Development," *Processes*, vol. 10, no. 9, pp. 1–25, 2022, doi: 10.3390/pr10091782.

[18] A. H. Ngu, M. Gutierrez, V. Metsis, S. Nepal, and Q. Z. Sheng, "IoT Middleware: A Survey on Issues and Enabling Technologies," *IEEE Internet Things J.*, vol. 4, no. 1, pp. 1–20, 2017, doi: 10.1109/JIOT.2016.2615180.

[19] J. Zhang, M. Ma, P. Wang, and X. Sun, "Middleware for the Internet of Things: A survey on requirements, enabling technologies, and solutions," *J. Syst. Archit.*, vol. 117, p. 102098, 2021, doi: 10.1016/j.sysarc.2021.102098.

[20] OGC, "Welcome to The Open Geospatial Consortium," *OGC.* https://www.ogc.org/ (accessed Aug. 10, 2023).

[21] OGC, "Open Geospatial Consortium OGC SensorThings API Part 1: Sensing," 2016. [Online]. Available: http://www.opengis.net/doc/is/sensorthings/1.0.

[22] M. Krommyda *et al.*, "An integrated toolbox for the engagement of citizens in the monitoring of water ecosystems," *Electron.*, vol. 9, no. 4, 2020, doi: 10.3390/electronics9040671.

[23] M. J. G. V. de Castro, "Data Models for Smart City IoT," University of Porto, 2017.

[24] T. Hardy *et al.*, "Sen2Grass: A Cloud-Based Solution to Generate Field-Specific Grassland Information Derived from Sentinel-2 Imagery," *AgriEngineering*, vol. 3, no. 1, pp. 118–137, 2021, doi: 10.3390/agriengineering3010008.

[25] Antonello Zanini, "JSON vs. JSONB in PostgreSQL: A Complete Comparison," *DbVisualizer*, 2022. https://www.dbvis.com/thetable/json-vs-jsonb-in-postgresql-a-complete-comparison/ (accessed Aug. 23, 2023).

[26] A. Chatterjee, M. W. Gerdes, P. Khatiwada, and A. Prinz, "SFTSDH: Applying Spring Security Framework with TSD-Based OAuth2 to Protect Microservice Architecture APIs," *IEEE Access*, vol. 10, pp. 41900–41920, 2022, doi: 10.1109/ACCESS.2022.3165548.

[27] A. B. Kretarta and H. Kabetta, "Secure User Management Gateway for Microservices Architecture APIs Using Keycloak on XYZ," *2022 5th Int.*

Semin. Res. Inf. Technol. Intell. Syst. ISRITI 2022, pp. 7–13, 2022, doi: 10.1109/ISRITI56927.2022.10052901.

[28] FRaunhofer, "FROST®-Server – An open source implementation of OGC SensorThings API," *GitHub*, 2023. https://github.com/FraunhoferIOSB/FROST-Server (accessed Aug. 14, 2023).

[29] StatusCake, "StatusCake: Free website uptime monitoring." https://www.statuscake.com/free-website-monitoring/ (accessed Aug. 14, 2023).

DOI: 10.1561/9781638282839.ch3

Chapter 3

AI for Corrosion Detection in Oil and Gas Refineries

By Leonidas Valavanis, Georgios Sarigiannis, Athanasios Karydis, Konstantinos Loupos and Ioanna Fergadiotou

As critical infrastructures continue to expand and play an essential role in modern society, the challenges associated with their inspection and maintenance become increasingly significant. Traditional methods often fall short in addressing the complexities of these infrastructures, which demand meticulous and efficient monitoring to ensure operational reliability, safety, and longevity. Thus, in the context of the EU-funded project PILOTING (No. 871542), we experiment with convolutional and transformer-based deep learning neural networks for the automated detection of corrosion on pipes and inside of vessels. The dataset used to train the networks is collected manually. To increase the dataset, augmentation methods such as Cycle-Gans [1] are employed. Experiments show that transformers outperform convolutional neural networks both in pipes and vessels and we can successfully detect corrosion on pipes and inside vessels with high Intersection over Union (IoU).

3.1 Introduction

Corrosion poses a significant challenge in industries and infrastructural domains where pipes and vessels are crucial components. Pipes and vessels are prone to corrosion, which can result in structural degradation, safety hazards, environmental concerns, and financial losses. Detecting corrosion in its early stages is essential to preventing its adverse effects. Traditional inspection techniques, which often are manual and time-consuming, have limitations when it comes to determining the level of corrosion on large building sites.

A new era of automated inspection and corrosion detection has been ushered in by developments in computer vision, artificial intelligence, and deep learning approaches. By harnessing the power of artificial intelligence and machine learning, this work aims to automate corrosion detection and improve the overall integrity of infrastructure. The urgency for accurate and efficient corrosion detection in construction sites stems from the potential ramifications of unchecked degradation. Corrosion not only compromises the structural integrity of pipes and vessels but also contributes to operational inefficiencies, maintenance costs, and environmental risks. Furthermore, manual inspection methods are often labour-intensive and subject to human error, which can result in missed or misdiagnosed corrosion instances. As construction projects continue to expand in scale and complexity, there is a pressing need for intelligent solutions that can rapidly and reliably identify corrosion anomalies.

In this work, we use a supervised object detection approach to train convolutional neural networks in order to automate detection of corrosion on pipes and inside vessels. Convolutional neural networks such as YOLO [7] and MaskR-CNN [4] have been used which have proven successful in detecting objects in different situations. Moreover, newer approaches such as Swin Transformers [6] have been used, which is a transformer-based deep neural network that uses self-attention and is state-of-the-art in object detection and many computer vision tasks. For the task of detection corrosion inside vessels we use the approaches as mentioned in the original papers. For the task of detecting corrosion on pipes we experiment with an extra object detection layer in order to mask out everything except pipes, which reduces false positives significantly.

The dataset is collected manually from different sources. The annotation of the images in the dataset is done using CVAT (computer vision annotation tool). After the initial annotations and training of the models, CVAT supports AI-assisted image annotation which annotates automatically images using trained models and the user can either accept or reject the automated annotations.

3.2 Methods

3.2.1 CycleGANs

The number of images needed to train a neural network is significant. Other than the standard augmentations like rotation, flipping, magnification, etc., a CycleGan [1] is used to generate images with corrosion in order to improve the training dataset and the performance of the model. CycleGANs [1] (Cycle-Consistent Adversarial Networks) are a type of generative deep learning model used for unpaired image-to-image translation. Images from Domain A are translated into Domain B by the model. This form of generative model has the benefit of not requiring matched samples of input and output for training, which speeds up the dataset creation process. There are many uses for CycleGANs, such as style transfer, domain adaptation, and image-to-image.

The CycleGan architecture comprises of the generator and discriminator networks. The generator, which uses an encoder-decoder model architecture, attempts to produce an output image in the other domain that corresponds to the input image. The discriminator network, in order to differentiate between the generated images and the actual images from the target domain, uses a deep convolutional neural network that performs image classification. Figure 3.1 shows the complete architecture of the generator and discriminator.

The generator and discriminator networks play a minimax game during training. The discriminator's job is to identify and categorize the created images as fake or real, whereas the generator's mission is to produce realistic images inside the target

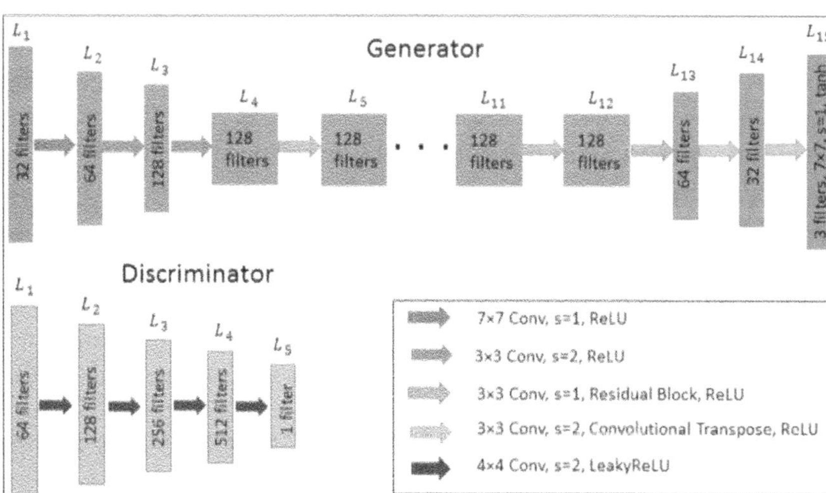

Figure 3.1. Architecture of the generator and discriminator of unpaired CycleGAN. Conv: 2D Convolutional filter, s: Stride, ReLU: Rectifier linear unit.

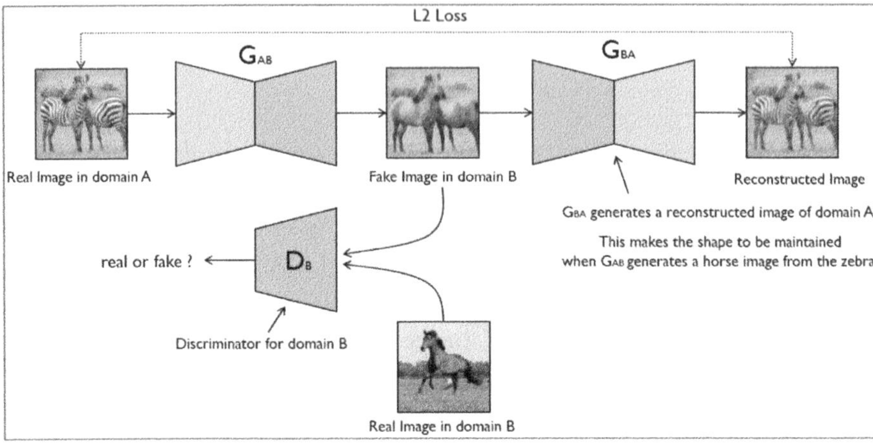

Figure 3.2. CycleGANs Architecture with example domains horse, zebra.

domain. The two networks' performance and knowledge are continually improved through this iterative process by mutually influencing one another, and eventually the generator develops the capacity to create convincing images that deceive the discriminator. Figure 3.2 shows the complete architecture of CycleGANs. CycleGANs use a loss function made up of the adversarial loss and the cycle consistency loss to train the generator and discriminator networks. Since the generator is trained to reduce overall loss, realistic images that are true to the source image can be produced. To discriminate between generated images and actual images from the target domain, the discriminator is trained to maximize the total loss.

In the loss function, the cycle consistency loss is a key component for the successful training. Cycle consistency ensures that there is minimal quality and information loss while translating the created image back to its native domain. To make sure of that, it uses a generator to translate the input image into the target domain, and then another generator to translate it back into the original domain. Forward cycle consistency refers to the translation to the target domain, and backward cycle consistency to the originating domain. The difference between the original input image and the image produced by the two-stage translation process is used to compute the loss.

Adversarial loss is employed to guarantee that the generated images are accurate and fall under the target domain. Based on the likelihood that the discriminator will properly classify the created images as fake, the loss for the generator is determined.

3.2.2 Corrosion Detection

In industrial settings, such as construction sites, corrosion detection is essential. Convolutional neural network-based deep learning models have demonstrated promising results in successfully and accurately recognizing corrosion. For instance, VGG [8] and ResNet [9], which are deep neural networks with numerous layers of convolutional and pooling operations, followed by fully connected layers for the final output predictions, can be utilized to detect corrosion. Transformer-based networks, which are state-of-the-art neural networks that employ self-attention methods to selectively attend to distinct parts of the input, have also become more popular due to their capacity to manage data, particularly in natural language processing. Transformers' main strength is that they can be taught to extract semantically significant characteristics from the input images in computer vision and in result corrosion detection. These features can then be given as input into a convolution neural network or a classification model for classification. In our solution, corrosion detection in tanks and pipes is done using supervised neural networks.

3.2.2.1 Corrosion detection on pipes

3.2.2.1.1 YOLO

YOLO ("You Only Look Once") [7], is a family of object detection real-time algorithms. Because they can perform well in real-time and have compact size models, YOLO models are popular. They are the best for real-time situations and on-device deployment scenarios because of their characteristics. YOLO is a single-stage object detector, which can predict every bounding box in a single model pass. It is quicker and more compact than a two-stage detector, which first suggests regions of interest before classifying the regions to objects.

There are three main parts to the YOLOv5 architecture, the *model backbone*, the *model neck* and finally the *model head.*

The **model backbone** identifies and extracts important features from an input image. Cross Stage Partial network (CSPNet) [2] is used as a backbone because it reduces computing bottlenecks, reduces memory costs and improves the capabilities of CNN. The authors claim that during network optimization, gradient duplication is what is responsible for the redundant calculations. By incorporating the gradient changes into the feature map from beginning to end, CSPNet minimizes the number of calculations required. A DenseNet and a CSPNet with integrated adjustments are shown in Figure 3.3. The shallow features are mapped twice using CSPNet. One half is concatenated directly with the output of the Partial Dense Block, and the other part is passed through the whole Dense module.

To help the model make generalizations about object scaling, it uses a **model neck**. Therefore, it is useful to recognize the same thing in various scales and sizes. A feature pyramid network is employed to do it. A Path Aggression Network, or PANet [3], is employed in YOLOv5. Because PANet reliably maintains spatial information, it can be used with YOLOv5. The main characteristics of PANet are

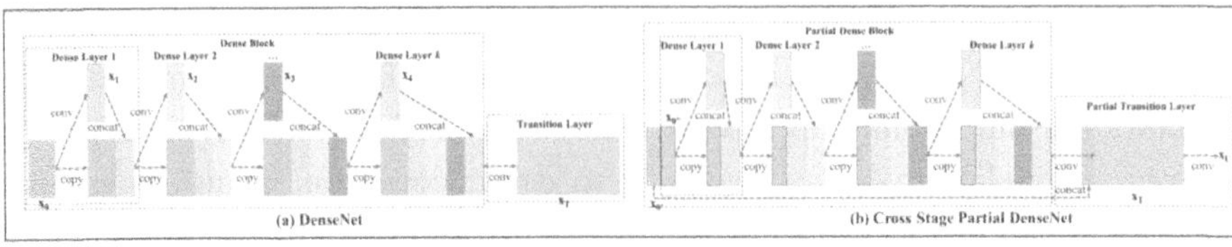

Figure 3.3. DenseNet vs CSPNet.

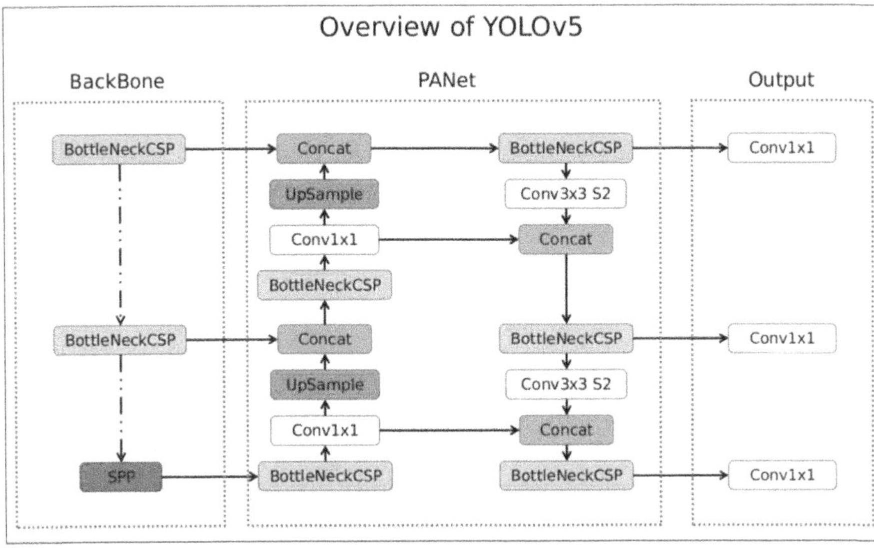

Figure 3.4. YOLOv5 architecture.

Bottom-up path augmentation, adaptive feature pooling and fully connected fusion to improve mask predictions.

The final detected objects are forecasted using a dense prediction process using the **model head**. The prediction consists of the predicted bounding box's center, height, and width, as well as the prediction's confidence score and the class's related name.

Figure 3.4 depicts a high-level picture of the full YOLOv5 architecture, with detailed descriptions of the head, neck, and backbone. YOLOv5 augments the data online in each batch for training purposes. Scaling, colour space changes, and mosaic augmentation are examples of data augmentation. Mosaic data augmentation, which turns four photos into four random-ratio tiles, is the most novel of them. The "small object problem," in which small items are harder to identify than larger ones, is addressed through mosaic augmentation.

3.2.2.1.2 MASKRCNN

Mask R-CNN [4] addresses the task of instance classification. The network can identify the various objects in an image and forecast the bounding boxes, classes, and masks. It is a development of Faster-RCNN [5], a region-based convolutional neural network that provides bounding boxes for each object and a confidence score for the class label that corresponds to it. An additional branch of the Mask R-CNN produces a segmentation mask for each bounding box. By using this segmentation mask, it solves the limitation of YOLO which just produces bounding boxes.

Mask R-CNN is a two-stage detector. In the first stage, a backbone network is used to analyse the image and identify regions of interest (RoI). Regions of interest are areas of an image that have an object in them. For each proposed region in Stage 1, the second stage forecasts the bounding boxes, object class, and segmentation mask.

Mask R-CNN architecture comprises of three main parts, the backbone, the Region Proposal Network (RPN) and finally the ROI, Mask classifier and bounding box regressor.

A convolutional neural network serves as the backbone and is employed as a feature extractor. ResNet is used as the **backbone**. A Feature Pyramid Network is utilized as an extension in the backbone to represent things at various scales.

The image is scanned by the **region proposal network**, which then suggests areas where the objects are. These scattered boxes in the image are referred to as anchors. In order to completely cover the image, anchors of various sizes and scales overlap. As opposed to processing the image directly, the network processes the features retrieved from the backbone fairly quickly. The bounding box refinement and anchor class are the RPN's outputs. Foreground and background are the two classes that make up the anchor class. The box in the foreground suggests that something is presumably inside of it. The bounding box refinement modifies the coordinates to better suit the box because the anchor might not be centered over the object. The suggested regions are filtered using non-max suppression, and the high probability regions are selected.

The suggested RPN regions are sent to a different network, where they are warped into a certain dimension. The classification and boundary box predictions are improved during this process. Finally, a mask classifier that outputs a binary mask for each RoI is given the warped features.

Figure 3.5 presents the full architecture of Mask RCNN in detail.

3.2.2.1.3 Swin transformers

Swin transformers [6] (shifted window transformers) is a state of the art transformer-based deep learning model that uses self-attention. It is suitable to be utilized as a basis for feature representation of images because it is very efficient and accurate in many computer vision tasks.

Swin Transformer's main concept is to divide the input image into non-overlapping patches and process them hierarchically utilizing a number of

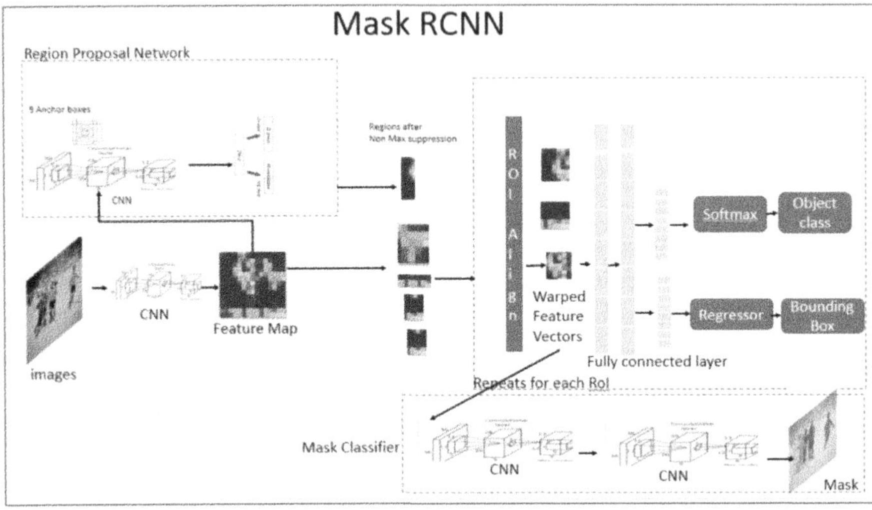

Figure 3.5. Mask R-CNN architecture.

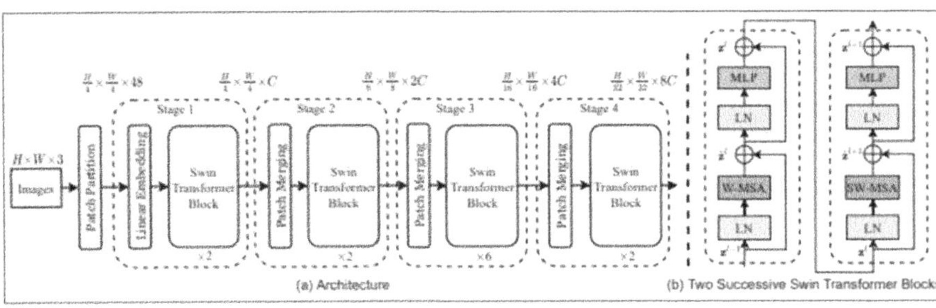

Figure 3.6. Swin Transformers architecture.

Transformer layers. By merging the patches into various stages utilizing patch merging layers, the patches are processed in a hierarchical manner. Similar to the original Transformer architecture, each stage is composed of a number of blocks, each of which has a set of self-attention layers followed by feed-forward layers. The complete architecture is presented in Figure 3.6.

There are two key concepts in Swin Transformers implementation, shifted window attention and hierarchical feature maps.

The *shifted window attention* makes it possible to process more varied spatial interactions between patches. Patches have no spatial relationship in a conventional transformer architecture because they are handled individually inside the same block of contrast, each block of Swin Transformer has a grid-organized collection of patches, and each patch is subjected to a number of self-attention and feed-forward layers for processing. Instead of processing every patch in a block simultaneously, each block's patches undergo a specific number of pixel shifts in both the horizontal and vertical directions.

The Swin Transformer network can effectively capture spatial interdependence at various scales thanks to its *hierarchical architecture*, which enables it to collect both local and global information from the input image. To do this, the model's receptive field is gradually expanded from small patches to the entire image. The input image is first divided into small patches using Swin Transformer, which then processes each patch using a succession of blocks with self-attention and feed-forward layers. Each stage that these blocks are organized into processes patches of a variable size. Different granularities of information are captured at each stage. Higher levels capture global information, whereas lower stages capture local information.

3.2.2.2 Corrosion detection inside tanks

Deep learning convolutional neural networks, which are capable of spotting objects, patterns, and abnormalities that could otherwise go unnoticed, can be used to identify corrosion inside tanks. Learning to spot patterns and irregularities in the images, including color discoloration, is a standard approach. By employing supervised algorithms to spot corrosion inside tanks, we were able to evaluate the advantages of the object detection models. More specifically, the MaskRCNN and Swin Transformer were both used in this instance, with the former serving as the model's backbone.

3.3 Dataset Synthesis

A crucial building ingredient in a neural network's learning process is the dataset. Neural networks pick up patterns and features from the dataset. The network can thus recognize and learn these patterns thanks to a well-curated dataset. To ensure that the model generalizes well to new and unseen data, the dataset must contain a variety of representative data. Additionally, the quantity and quality of the data in a dataset might affect how well a neural network performs. The accuracy and performance of the model can be enhanced with a larger dataset and more diverse instances, which will enhance performance on the given task. Finally, there should be no bias in the dataset toward particular sample types. To prevent bias and guarantee accurate predictions, it is crucial to select a dataset that is indicative of the actual problem. When considering the detection of corrosion in pipes, where the background exhibits significant variance, the significance of the dataset is substantially increased. More specifically, it is stated in [10] that more than 65,000 annotated photos are required to achieve an artificial intelligence model's ability to identify corrosion and segment it (i.e., label each pixel).

Collecting a dataset for corrosion has proven a challenging task. The CHEVRON site could not be accessed because of COVID restrictions, and there aren't many free datasets that are appropriate for this particular use case. An initial batch of 330 photos was acquired through the network of CATEC partners. This collection consists of pipe-related photos that show both corrosion-free and corroded sections. Moreover, 1223 additional photos were included in the collection thanks to INLECOM's registration in SPRINT robotics. The initial dataset for pipes, designated Dataset A, is made up of these two batches of photographs. The total number of photos in this collection is 1553, of which 209 are defective and the remainder are healthy. It has 3993 corrosion annotations and 2845 pipe annotations.

Six distinct drone missions from the Chevron expedition yielded a total of 1680 photos. After carefully examining the photos, it was discovered that some of the images show overlapping zones, and occasionally, different missions travel over the same areas. Additionally, several photographs show pipes and other items at large distances, which differs from a realistic depiction of real-world situations. The decision was made to eliminate these photographs because keeping them would lengthen the annotation process and provide no benefit for teaching the algorithms to recognize corrosion in the images. 526 of the 1680 photos were kept out of the total. The second dataset, designated Dataset B, was produced by included these 526 photos in the initial dataset. The pipes dataset includes 2079 photos in total, 457 of which are healthy, as well as 26158 corrosion annotations and 6138 pipe annotations. The dataset is suitable for training a robust model since the photos in it feature different angles of illumination of the scenery. These pictures demonstrate pipe corrosion on building sites.

Finding photographs of tanks other than the Pilots proved challenging due to the task's extreme specificity. As a consequence, 900 photos in all were acquired together with 1377 annotations. Figures 3.7 and 3.8 illustrate sample images taken during the pilots for pipes and tanks respectively.

Data augmentation strategies were investigated to improve the models' performance and robustness. CycleGANs were specifically designed to produce images with corrosion. Approximately 70% of the data were utilized for training, 20% for validation, and 10% for testing the corresponding models. During the training and validation phases, the network is not shown any of the test images. Table 3.1 below provides a thorough summary of each dataset, including the total number of photos, healthy images, annotations, and splits. Subsequently, Table 3.2 presents the total number of images gathered per task.

The pipe's dataset has a large number of corrosion annotations because every instance of corrosion, whether or not it is on a pipe, was noted in the image. The successful training of a neural network intended for corrosion identification depends heavily on this factor. It is crucial that every incidence of corrosion in the picture collection is precisely tagged in order for the model to develop a thorough

Figure 3.7. Sample image taken during the pilots for detecting corrosion in pipes.

Figure 3.8. Sample image taken during the pilots for detecting corrosion.

Table 3.1. Datasets description NA=Not available.

	# Images	# Corrosion annotations	# Pipes annotations	# Healthy images	# Train images	# Validation images	# Test images
Pipes DatasetA	1553	3993	2845	209	1080	310	163
Pipes DatasetB	2079	26158	6138	457	1455	415	209
Tanks	900	1377	NA	27	630	180	90

Table 3.2. Total number of images gathered for each task.

	# Images
Pipes	2079
Tanks	900

Table 3.3. Enhanced datasets using CycleGANs.

	# Images	# Corrosion annotations	# Pipes annotations	# Healthy images	# Train images	# Validation images	# Test images
Pipes DatasetC	2379	31023	6352	457	1705	465	209
CycleGANs	315	NA	NA	151	220	63	32

grasp of corrosion. As a result, the annotations need to be as accurate and thorough as feasible. However, as this is the sole set being evaluated, only the corrosion on pipes is marked in the test set.

CycleGANs

The first dataset for training CycleGANs comprises of 130 images (81 with corrosion and 34 without corrosion). Pipes' complex background makes it challenging for generative models to produce the high-quality images required to train AI algorithms. The dataset was therefore expanded as a result of the complex background and class imbalance. The difficult background was removed from the photos, further crops of images without corrosion were added, and the dataset was increased to 315 images (154 with corrosion and 151 healthy).

To increase the dataset for pipes, 300 pictures were produced using the CycleGAN generating network. To prevent bias during training, only 300 images were added to the initial dataset. The final collection, designated collection C, consists of 2379 photos with 6352 pipe annotations and 31023 corrosion annotations. The datasets are presented in Table 3.3.

3.4 Results

Validation process and Quality Assessment are important aspects in Deep Learning, a field that involves training and developing models to perform various tasks. The validation process and quality assessment involve evaluating trained models to unseen data to see how well the model can generalize and avoid overfitting and also evaluate on various performance metrics depending on the task. In the sections below, we present, for each task, the corresponding validation metrics used as well as the experiments conducted and their results. For all experiments an AWS instance was used with a V100 Nvidia GPU with 16 GB of memory.

3.4.1 CycleGANs

It can be challenging to compare the performance of numerous trainings with various hyper-parameters because there is no objective loss function during training for generative networks. Therefore, both a qualitative and a quantitative evaluation of a generative network must be done. The resulting photos are compared and visually examined for qualitative evaluation. The Fréchet Inception Distance (FID) metric is used for quantitative evaluation. By comparing the distributions of the original images and the generated images, FID measures the quality of the generated images. It makes use of the Fréchet distance, which calculates how similar two probability distributions are to one another. The probability distributions represent the distribution of features extracted using a trained Inception-v3 neural network for both the real and generated images.

When the features of the real and created images are more similar the FID is lower. A lower FID score denotes a higher quality of the generated images. As it considers both the quality and diversity of the generated images, FID is a more trustworthy metric than others, such Inception Score. Figure 3.9 illustrates how an image's distortion and FID correlate with one another [11]. As demonstrated, an image with a FID greater than 100 exhibits noticeable distortion.

The FID metric is not ideal for this task since it only captures the marginal distribution between the input and generated images and ignores domain translation, or the alignment of the output and input. To compare the generated images with a ground truth image, a more acceptable metric would be to have matched images of corrosion and healthy images. Collecting such a dataset is not feasible due to our restricted resources and the specialized objective of detecting corrosion.

The dataset, which consists of 63 images for validation and 220 images for training, is presented in the Dataset Synthesis section. At first, there were 130 total photos in the dataset. Due to an unbalanced sample and a diverse context, the results obtained utilizing the first dataset were unsatisfactory. Learning the distribution of the photos is challenging when the background of the photographs is complicated and variable. As a result, low variance photos were used to replace the original dataset. Additionally, we constructed cropped photos from healthy portions of other images (not included in the dataset) to improve the dataset and include additional healthy images to correct the imbalance. Since the dataset of the pre-trained models was irrelevant to our use case and would not be useful for the training, no

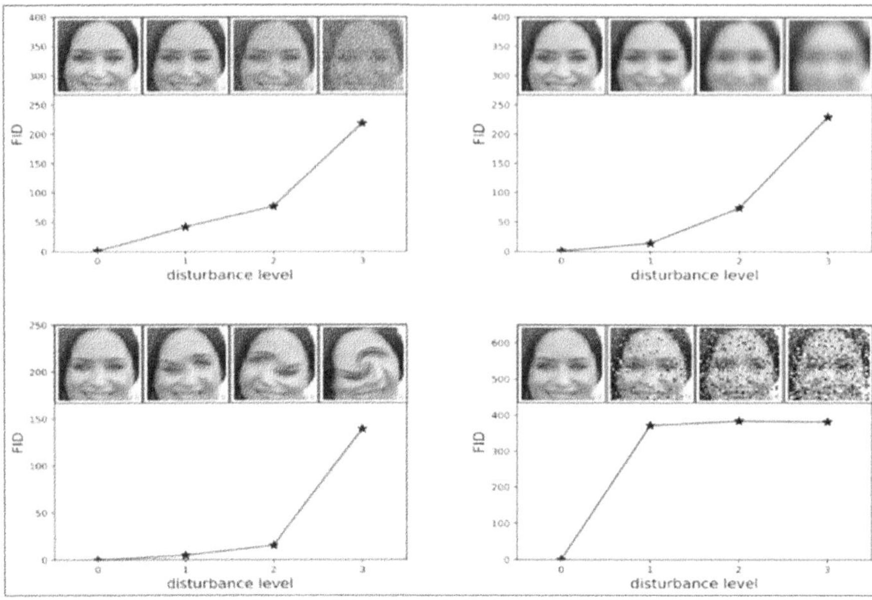

Figure 3.9. Example of How Increased Distortion of an Image Correlates with High FID Score.

Figure 3.10. Sample images used for training CycleGANs. Left image: Image with low variance, Right Image: Augmented healthy image created by using crops from other images.

pre-trained model was employed to establish the weights. Figure 3.10 shows a low variance image (left) and an enhanced image that was made by utilizing portions of other photographs.

The input size for the photos in our experiments is 256 × 256. We tested a range of batch size, learning rate, and input size (128 × 128) values. To prevent overfitting, the models were trained for 200 iterations, and the best model with the lowest validation loss was selected. In order to prevent overfitting, flipping, rotation, and cropping augmentation techniques are also used during training. The outcomes

Table 3.4. Results of quantitative evaluation using CycleGan.

Images input size	Batch size	Learning rate	FID on test set
128 × 128	10	0.0002	68.26
128 × 128	10	0.001	77.31
128 × 128	8	0.0002	71.46
256 × 256	8	0.0002	55.13
256 × 256	10	0.001	51.08
256 × 256	10	0.0002	**42.19**

Figure 3.11. Generated sample using best model from experiments. (Left image): Input image to GAN, (Right Image): Generated image.

of the experiments employing CycleGANs are shown in Table 3.4. The best results are obtained with a FID score of **42.19** while using 256 × 256 photos, a batch size of 10, and a learning rate of 0.0002. According to the aforementioned research [11], the generative model is able to capture the distribution of the images based on the FID score of 42.19. Additionally, it is not low enough to suggest that images are identical. As a result, technique may successfully transform healthy photos into corroded images.

We created 300 photos for the dataset using the best performing model from the experiments. Figures 3.11 and 3.12 show the generated images along with their original input photos to provide a qualitative evaluation. As demonstrated, the model is able to both create photos with corrosion and capture the distribution of corrosion images. The checkerboard pattern is an artifact in the output photos, despite the fact that they are consistent with the original input image. This happens when the input data is upsampled to create the image, which happens rather

Figure 3.12. Generated sample using best model from experiments. Left image: Input image to GAN, Right Image: Generated image.

frequently when training generative models. We tried training for more epochs and making the discriminator's receptive field bigger, but the results were similar.

3.4.2 Corrosion Detection on Pipes

Pixel-based defect identification systems are evaluated according to their ability to classify corrosion instances with an IoU of at least 70%. The IoU was used to compare the bounding boxes of the ground truth and the prediction in order to assess the performance of object detection systems. We determined the amount of overlap between the ground truth and forecasted bounding boxes and divided it by the areas of union to determine the IoU. We also averaged each image's incidences of IoU corrosion, followed by an average across all images. We tested different approaches for detecting corrosion on pipes while continuously adding images to our collection. We tested multiple algorithms and techniques to get the best outcomes feasible due to the task's intricacy. Thus, the validation process was applied to each model, and the results are displayed below.

Following the authors' advice,[1] we trained the model using YOLO. In order to achieve label consistency, which requires that every instance be labelled in every image, we employed 1500 photos per class, at least 10,000 annotations per class, image variety for diversity, and at least 10,000 annotations per class. The input photos to the network are 640 × 640 in size.

The learning rate and the likelihood of using the mosaic augmentation on the input images were tested. In the mosaic augmentation method, four separate images

1. https://docs.ultralytics.com/yolov5/tutorials/tips_for_best_training_results/.

Table 3.5. Results using YOLO.

Dataset	Learning rate	Mosaic augmentation	IoU on test set (%)
DatasetA	0.00258	1	45,52
DatasetA	0.001	1	43,36
DatasetA	0.001	0.8	49,11
DatasetA	0.0001	1	47,66
DatasetB	0.001	0.8	58,83
DatasetB	0.0001	1	60,24
DatasetC	0.001	0.8	59,28
DatasetC	0.0001	1	**61,54**

Figure 3.13. Example image using YOLO best model.

are combined into one and then subjected to various transformations, including scaling, rotation, and flipping. To prevent overfitting, the models were trained for 150 epochs, and the best model with the lowest validation loss was selected. On the COCO dataset, which consists of 140,000 images and 80 classes, we employed a pre-trained model. As a result, the model's weights have already been initialized with general features, and given our dataset, it will focus on detecting corrosion.

Table 3.5 shows the outcomes while employing YOLO. On the test set, the best model produces **61,54%** IoU. It has been demonstrated that performance rises along with dataset size. Additionally, using the CycleGANs-generated data improves the results. According to the authors' article, YOLO struggles to locate small items and has some spatial limits that limit the number of objects it can predict. Furthermore, because to the architecture's numerous downsampling stages, it has trouble generalizing to objects with various aspect ratios and configurations. Since corrosion can take many different forms, our dataset contains many annotations with small and irregularly shaped examples.

Figure 3.13 illustrates an example where corrosion is predicted. As shown, confidence is low and small objects are difficult to identify.

As we continued to add more images to our dataset, we conducted experiments using the three datasets indicated in the dataset synthesis section before moving on to the evaluation of the MaskRCNN model. In order to maintain the original aspect ratio, the input images to the network are 1024 × 1024 padded. In order to normalize photos using those values, the mean pixel of the dataset is also determined for each of the three channel colours (Red, Green, and Blue). Two images of 1024 × 1024 could fit on the GPU, hence the batch size was 2.

We experimented with the learning rate, weight decay, minimal confidence of the model, and backbone model. Predictions with less confidence than the value during training will be discarded by the minimum confidence. During training, the maximum number of detections is also utilized to regulate the quantity of detections. To prevent overfitting, the models were trained for 100 epochs, and the best model with the lowest validation loss was selected. The COCO dataset was used with a pre-trained model. In order to prevent overfitting, flipping, rotation, cropping, and blurring augmentation techniques are used during training.

The outcomes of training MaskRCNN to find corrosion on pipes are shown in Table 3.6. On the test set, the best model produces an IoU of **77,29%**. A smaller number of experiments are shown because there are many hyper-parameters. As backbone models, we tested with ResNet50, ResNet101, Swin-T, and Swin-S. Resnet50 and Swin-T perform better than their more powerful alternatives, Resnet101 and Swin-S, as seen by the results. Because the dataset was tiny and a complex model would have overfitted and been unable to generalize to new data, a simpler model with fewer parameters performed better. Regarding the quantity of

Table 3.6. Results in test set for detecting corrosion on pipes using MaskRCNN.

Dataset	Backbone model	Min confidence	Weight decay	Learning rate	IoU on test set (%)
DatasetA	Resnet50	0.5	0.0001	0.01	62,38
DatasetB	Resnet101	0.5	0.0001	0.001	64,03
DatasetB	Resnet101	0	0.0001	0.01	66,71
DatasetB	Resnet50	0	0.0001	0.001	71,25
DatasetB	Resnet50	0	0.0001	0.01	72,59
DatasetB	Swin-S	0	0.0001	0.01	74,61
DatasetB	Swin-T	0	0.0001	0.01	76,43
DatasetC	Resnet101	0	0.0001	0.01	67,82
DatasetC	Resnet50	0	0.0001	0.01	72,91
DatasetC	Swin-S	0	0.0001	0.01	73,11
DatasetC	Swin-T	0	0.0001	0.01	**77,29**

Figure 3.14. Sample image predictions on the test set using MaskRCNN. Green is the ground truth annotation, red is the prediction. Next to each annotation there is the confidence/IoU.

parameters and the size of the model, Swin-T and Swin-S are equivalent to Resnet50 and Swin-S, respectively.

Examples of some example photographs with predictions from the best model are shown in Figures 3.14 and 3.15. In the instances, we highlight the ground truth annotation in green, the prediction in red, and above each prediction, the model's confidence and IoU (in the format confidence/IoU). The model can successfully and with high confidence identify corrosion in these photos despite the varying lightning conditions and distance from the pipes.

Without doing any pre-processing, we examined the MaskRCNN findings and noticed that we occasionally make false positive predictions, which is to be expected given that the backdrop varies. We therefore used a two-step method for the end result to try and solve this issue. Since we have the annotations for the pipes, we use MaskRCNN to identify the pipes in the first stage, mask the rest of the image,

Figure 3.15. Sample image predictions on the test set using MaskRCNN. Green is the ground truth annotation, red is the prediction. Next to each annotation there is the confidence/IoU.

and then feed the masked image to the corrosion model to estimate the corrosion on the pipes. Since everything else in the image is hidden, this technique eliminates false positives while also making it simpler for the model to detect corrosion. For inference, we use MaskRCNN twice, once to predict the pipes and subsequently the corrosion in the masked image. MaskRCNN is trained to recognize pipes and corrosion in the same network.

The settings and experiments follow the pattern shown above. The outcomes with MaskRCNN masked are shown in Table 3.7. We only tested this approach using Datasets B and C and the background of the Swin transformers. On the test set, the best model produces an IoU of **71,46%**.

This technique is dependent on the pipes' preliminary prediction. Therefore, if the pre-processing stage is unsuccessful, the corrosion detection will also be

Table 3.7. Results in test set for detecting corrosion on pipes using MaskRCNN masked.

Dataset	Backbone model	Min confidence	Weight decay	Learning rate	IoU on test set (%)
DatasetB	Swin-S	0	0.0001	0.01	69,42
DatasetB	Swin-T	0	0.0001	0.01	71,12
DatasetC	Swin-S	0	0.0001	0.01	70,57
DatasetC	Swin-T	0	0.0001	0.01	**71,46**

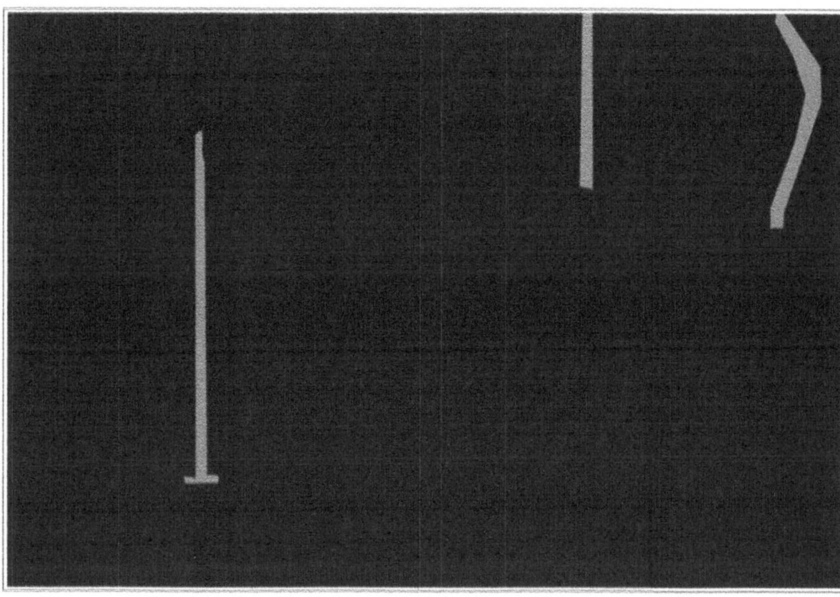

Figure 3.16. Example masking of pipes for image.

unsuccessful. An illustration of hiding the pipes is shown in Figure 3.16. The findings obtained using the best model are shown in Figures 3.17 and 3.18. As evidenced by the data, some pipes with corrosion go undetected because to improper pipe masking, but the number of false positives has greatly decreased.

3.4.3 Corrosion Detection Inside Tanks

As it was mentioned in the previous section, the metric for detecting corrosion inside tanks is IoU > 70%. Due to the similar and simpler background, identifying corrosion in tanks is easier than detecting corrosion on pipelines. Therefore, we only tested utilizing MaskRCNN with no input image pre-processing. Furthermore, as demonstrated in the detection of corrosion in pipes, MaskRCNN predicts on a

Figure 3.17. Sample image predictions on the test set using MaskRCNN Masked. As shown, some corrosion on pipes is not identified due to the pipe not detected. Green is the ground truth annotation, red is the prediction. Next to each annotation there is the confidence/IoU.

pixel-level, making its predictions more accurate, whereas YOLO predicts bounding boxes and has a lesser confidence.

Experiments were conducted in which 630 photos were used for training and 180 were used for validation utilizing the dataset outlined in the dataset synthesis section. As previously mentioned for MaskRCNN on pipes, the normalization, data augmentation, picture input size to the network, and hyper-parameters remained the same. The COCO dataset was used to initialize the weights, and the model underwent 100 iterations of training. It was decided to use the model with the smallest validation loss. The outcomes of teaching MaskRCNN to recognize corrosion are shown in Table 3.8. On the test set, the best model produces an IoU of 89,18%. A smaller number of experiments are shown because there are many hyper-parameters. As backbone models, we tested with ResNet50, ResNet101, Swin-T, and Swin-S. Resnet50 and Swin-T perform better than their larger alternatives, Resnet101 and Swin-S, as seen by the results. Because the dataset is tiny and a complicated model would overfit and be unable to generalize to new data, simpler models with fewer parameters perform better.

Figure 3.18. Sample image predictions on the test set using MaskRCNN Masked. Green is the ground truth annotation, red is the prediction. Next to each annotation there is the confidence/IoU.

Table 3.8. Results for MaskRCNN.

Backbone model	Min confidence	Weight decay	Learning rate	IoU on test set (%)
Resnet101	0	0.0001	0.001	65,27
Resnet101	0.5	0.0001	0.001	62,30
Resnet101	0	0.001	0.001	61,33
Resnet101	0	0.0001	0.01	67,65
Resnet50	0	0.0001	0.001	82,65
Resnet50	0	0.0001	0.01	84,85
SwinS	0	0.0001	0.01	86,34
SwinT	0	0.0001	0.01	**89,18**

Figures 3.19 and 3.20 are a few test set examples of corrosion inside tanks predicted by our best model. We outline the ground truth annotation in green, the prediction in red, and above each prediction, the model's confidence and the IoU (in the format confidence/IoU), just as we did in the experiments of the preceding sections.

Figure 3.19. Sample image predictions on the test set using MaskRCNN. Green is the ground truth annotation, red is the prediction. Next to each annotation there is the confidence/IoU.

Figure 3.20. Sample image predictions on the test set using MaskRCNN. Green is the ground truth annotation, red is the prediction. Next to each annotation there is the confidence/IoU.

3.5 Conclusions

In conclusion, the use of machine learning algorithms for detecting corrosion is a promising area of research with significant potential for reducing maintenance costs. By leveraging machine learning algorithms, AI-based systems can analyse vast amounts of data collected from multiple sources to identify indications of corrosion damage. This can help maintenance teams to detect corrosion early on and take timely corrective actions, thereby reducing the risk of catastrophic failure and extending the lifespan of assets.

Despite the challenges with collecting the dataset, state-of-the-art methods were implemented and trained to identify corrosion on pipes and inside tanks surpassing 70% IoU. Regardless of the potential benefits of corrosion detection using AI models, there are still some challenges to be addressed, such as data quality and accuracy, algorithm selection and optimization as well as collecting data. However, with continued research and development, the performance of the algorithms can improve.

Acknowledgements

This research was funded by the PILOTING project under grant number 871542, supported by the European Union's Horizon 2020 innovation action program.

References

[1] J. -Y. Zhu, T. Park, P. Isola and A. A. Efros, "Unpaired Image-to-Image Translation Using Cycle-Consistent Adversarial Networks," *2017 IEEE International Conference on Computer Vision (ICCV), Venice, Italy*, 2017, pp. 2242–2251, doi: 10.1109/ICCV.2017.244.

[2] Wang, Chien-Yao *et al.* "CSPNet: A New Backbone that can Enhance Learning Capability of CNN." *2020 IEEE/CVF Conference on Computer Vision and Pattern Recognition Workshops (CVPRW)* (2020): 1571–1580.

[3] S. Liu, L. Qi, H. Qin, J. Shi and J. Jia, "Path Aggregation Network for Instance Segmentation," *2018 IEEE/CVF Conference on Computer Vision and Pattern Recognition*, 2018, pp. 8759–8768, doi: 10.1109/CVPR.2018.00913.

[4] K. He, G. Gkioxari, P. Dollár and R. Girshick, "Mask R-CNN," *2017 IEEE International Conference on Computer Vision (ICCV)*, 2017, pp. 2980–2988, doi: 10.1109/ICCV.2017.322.

[5] Shaoqing Ren, Kaiming He, Ross Girshick, and Jian Sun. 2015. "Faster R-CNN: towards real-time object detection with region proposal networks." In *Proceedings of the 28th International Conference on Neural Information Processing Systems – Volume 1 (NIPS'15)*. MIT Press, Cambridge, MA, USA, 91–99.

[6] Liu, Ze, *et al.* "Swin transformer: Hierarchical vision transformer using shifted windows." *Proceedings of the IEEE/CVF International Conference on Computer Vision*. 2021.

[7] Redmon, Joseph, *et al.* "You only look once: Unified, real-time object detection." *Proceedings of the IEEE Conference on Computer Vision and Pattern Recognition*. 2016.

[8] Simonyan, K., and Zisserman, A. (2014). Very deep convolutional networks for large-scale image recognition. arXiv preprint arXiv:1409.1556.

[9] He, Kaiming *et al.* "Deep Residual Learning for Image Recognition." *2016 IEEE Conference on Computer Vision and Pattern Recognition (CVPR)* (2015): 770–778.

[10] Nash, W., Drummond, T., and Birbilis, N. (2019). "Deep learning AI for corrosion detection." In *Corrosion 2019* (NACE – International Corrosion Conference Series). NACE International.

[11] Heusel, Martin *et al.* "GANs Trained by a Two Time-Scale Update Rule Converge to a Local Nash Equilibrium." *NIPS* (2017).

DOI: 10.1561/9781638282839.ch4

Chapter 4

Motorway Tunnels Structural Deformation Assessment Using Laser Scanning

By Andreas El Saer, Konstantinos Loupos, George Sarigiannis, Valia Konstantopoulou, Athanasios Karydis and Gerasimos Kouloumbis

In the current era where the majority of road infrastructures across the globe are grappling with severe aging, they continue to play a pivotal role in ensuring the smooth flow of transportation for both people and cargo. However, it is important to force monitoring and infrastructure attention, considering the challenges posed by climate and weather fluctuations, encompassing extreme events, natural and man-made disasters, as well as the gradual wear and tear of materials. Simultaneously, the growing demands on capacity due to increased user traffic necessitate our scrutiny. These factors collectively contribute to the overarching resilience of the road network, as a critical determinant of its long-term sustainability. The sustainability of this network reverberates throughout various facets of society, encompassing economic, social, and environmental dimensions, all of which underscore the intrinsic value of resilience. Consequently, there is an urgent call for measures such as adaptation strategies, monitoring, risk management and analysis, and advanced forecasting. Furthermore, there is a pressing need for effective vulnerability modeling and assessment techniques to fortify these critical lifelines.

Within this framework, this publication presents recent research findings focused on a pioneering 3D modeling approach and methodology designed to gauge deformations in highway tunnels. This innovative method employs 3D laser imaging to capture data and consolidates these results into a system geared towards the early detection of deformations and structural deteriorations that occur over time. This advanced system is conducts a comprehensive analysis of the infrastructure, achieving precision down to the millimeter level. It then compares this data with pre-existing models to pinpoint any anomalies in the 3D shape of the tunnel, particularly focusing on the intrados, which may manifest as partial deformations. This proactive approach provides an early indication of infrastructure deterioration, accurately georeferenced for immediate incorporation into the maintenance planning activities of highway safety teams and maintenance personnel. The system also offers invaluable 3D geometrical information pinpointing the exact location of defects. In the subsequent phase of development, the system can be seamlessly integrated into an autonomous navigating robotic platform, capable of positioning the sensing device at specific inspection points of interest or strategically chosen locations for holistic infrastructure measurement and modeling. This research represents the ongoing endeavors of the PILOTING project, currently in its second year of execution, and co-funded by the European Commission. The project focuses on robotic inspection solutions for critical infrastructures, encompassing both oil and gas facilities as well as transport infrastructures, including viaducts and tunnels. For more information on the project, please visit https://piloting-project.eu/.

4.1 Introduction

4.1.1 Ageing Infrastructures

In a European and Worldwide era where road infrastructures are significantly aging, their vital role in maintaining seamless and intermittent transportation for both people and cargo remains unquestionable. Given the challenges posed by climate change, extreme weather events, natural and human-made disasters, and the aging of structural materials, a strong focus on monitoring becomes important and should not be overseen. In parallel, the rising demand for increased transport capacity due to increased user flows further imposes needs for comprehensive attention over the structural capability of road infrastructures. Such factors constitute the needs on resilience of the entire road/transport network, an essential facet of ensuring its long-term viability and initial origin. As the sustainability of road networks directly impacts economic, social, and environmental dimensions, it highlights the value of holistic resilience. This, in turn, brings the urgency for adaptation strategies, monitoring, risk analysis and management, forecasting, and adoption of vulnerability assessment techniques while continuous monitoring and assessment is highlighted.

Getting deeper to infrastructures' recent challenges, aging infrastructure presents many pressing problems that need urgent attention and strategic solutions. Firstly, deteriorating structures (road, tunnels, bridges) pose significant safety risks to the users and public. The gradual degradation of materials and components compromises the structural integrity, leading to catastrophic failures. Outdated infrastructure, on top, hampers economic growth and efficiency. Inefficient transportation networks, negatively contribute to higher operational costs and reduced productivity and safety. As infrastructures age, maintenance and repair costs come to life. Financial constraints often force governments and public organizations to prioritize short-term fixes over comprehensive overhauls, imposing high risks in the cycle of degradation. Effective and continuous monitoring is crucial for identifying vulnerabilities in aging infrastructure, early enough. Traditional methods often fall short in providing real-time data on structural health, potentially leading to unexpected failures. Modern sensor technologies and remote monitoring systems can offer more accurate and timely insights, allowing for proactive maintenance and risk mitigation. Implementing these monitoring technologies, however, presents challenges. Integration with legacy systems can be complex and costly. Additionally, data security and privacy concerns arise when connecting infrastructure to the digital realm. A balance between connectivity and safeguarding critical information becomes essential.

Decision making is another important point strongly linked to predictive maintenance. Data and continuous monitoring build the necessary monitoring layer that is studied in each case. Without comprehensive data on infrastructure performance, it's difficult to allocate resources effectively and prioritize projects. A lack of clear visibility into asset conditions can lead to missed opportunities for optimizing performance and extending the lifespan of key infrastructure.

Within the above context, this publication presents recent research findings centered on a pioneering 3D modeling methodology. The approach employs cutting-edge 3D laser imaging sensors to gauge deformations within highway tunnels, contributing to early detection of structural changes over time. Operating at an extraordinarily precise (millimeter) level, the system analyzes a 3D representation of the infrastructure. This analysis is then juxtaposed with previous models to discern any anomalies in the tunnel's 3D shape, specifically within its intrados, which can emerge as a consequence of partial deformations. The system promptly identifies preliminary signs of infrastructure deterioration, accurately georeferencing these instances. This georeferenced data is then swiftly assimilated into the maintenance plans of highway safety teams and maintenance personnel. Additionally, it furnishes comprehensive 3D geometrical insights regarding the precise location of defects.

side-pavement, ventilation, drainage and other, it becomes important to not only acknowledge their unique characteristics but also address the specific challenges they pose. Despite the formidable obstacles associated with devising suitable approaches and methodologies for inspecting and maintaining tunnels, the consistent evaluation of their condition and the proactive response to potential damages remain paramount. This holds true for transportation authorities, inspection teams, and highway operators alike. Conducting inspections and maintenance activities within tunnels introduces significantly greater risks compared to working in open spaces, primarily due to their inherent attributes. Factors such as reduced lighting, shaded areas, elevated humidity levels, dusty and slippery conditions, and physical space constraints like narrow widths and obstructions contribute to the complexity. Moreover, the presence of live traffic adds an extra layer of complexity to the equation. Consequently, tunnel inspections must be executed promptly and accurately, demanding increased speed, cost efficiency, minimized personnel exposure, and the delivery of precise outcomes that are consistently updated. This approach allows for the comparison of damage progression over time, aiding in the identification of potential issues and their escalation trends [2].

4.2.2 Tunnel Inspection - Baseline Requirements

Baseline requirements in tunnel inspection include several important points that can be unarguable over safety, financial and other reasons as follows:

1. **Addressing Live Traffic:** In order to navigate the challenges posed by ongoing traffic and to ensure safety, a strategic approach was adopted. This involved temporarily halting traffic flow in one lane of the tunnel, either left or right, facilitating the precise placement of the laser scanner directly on the road pavement. This positioning enabled the acquisition of accurate and steady measurements for the creation of detailed 3D point representations.
2. **Streamlining Model Co-Registration:** The uniform design of tunnels, resulting in a consistent visual layout even across distinct sections, presented a challenge in manually aligning models. Given the ongoing nature of this task, an efficient solution appeared to lie in the implementation of tags. This approach is proving optimal for the co-registration process, significantly mitigating the time-intensive nature of the task.
3. **Varied Lighting Conditions:** The tunnel environment exhibits considerable fluctuations in lighting conditions, particularly at entry and exit points. Nonetheless, the decision to employ laser technology was astute, as it remains unaffected by the disparate lighting conditions prevalent within the tunnel. This choice has proven beneficial in ensuring reliable and consistent data collection.

4. **Establishing a Unified Coordinate System:** To accommodate the imperative of revisiting scanning locations with precision, the establishment of a standardized coordinate system emerged as a necessity. As this aspect remains a work in progress, the strategy leans towards leveraging an autonomous approach. This endeavor aims to guarantee accurate and consistent point referencing, enhancing the utility of the collected data for ongoing analysis and comparisons.

4.3 Laser Scanning for Tunnel Deformation Detection

As an integral facet of the comprehensive inspection process, the employment of the FARO scanner (Figure 4.1) played a pivotal role in the tunnel experiments. Its purpose was to execute highly precise 3D scans, rendering defects with exceptional accuracy at the millimeter scale. While the scan activities pertained to localized tunnel assessments, the trials encompassed both the road and drainage tunnels. These endeavors exclusively employed a vehicle robot due to the scanner's substantial weight and dimensions, making drone transportation unfeasible.

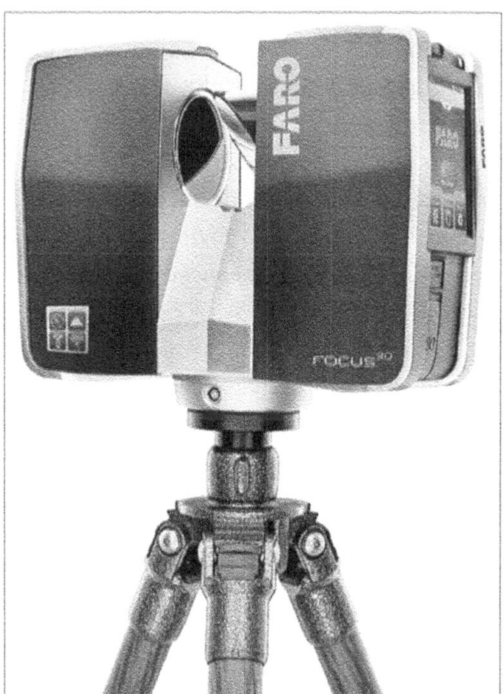

Figure 4.1. FARO 3D laser scanner used.

The FARO 3D laser scanner is a laser scanning device produced by FARO Technologies, a company specializing in measurement and imaging solutions. A 3D laser scanner is a sophisticated instrument that uses laser technology to capture detailed and accurate three-dimensional representations of objects, environments, or structures. It does this by emitting laser beams that bounce off surfaces and return to the scanner, allowing it to measure distances and create a precise digital model of the scanned area. FARO's 3D laser scanners are commonly used in various industries such as architecture, engineering, construction, manufacturing, and archaeology, among others. These scanners are particularly valuable for tasks such as quality control, reverse engineering, building information modeling (BIM), site documentation, and forensic analysis. They can capture intricate details, surface geometries, and complex structures that are challenging to measure using traditional methods. FARO offers a range of 3D laser scanners with varying specifications and capabilities to cater to different application needs. These scanners can be used for capturing data both indoors and outdoors, and they come with different levels of accuracy, range, and scanning speeds. The captured data is often processed using specialized software to create digital models, point clouds, and other visual representations that can be analyzed, measured, and manipulated for various purposes.

Regarding the particular scanner's resolution configuration, a deliberation led to the adoption of a 43.7-megapixel/3x quality setting. This decision was made to yield a cloud point-to-point accuracy of 6.1 mm at a 10-meter range. Given the proximity of 1 meter from the tunnel wall, the actual system accuracy in relation to the targeted lane side amounted to a commendable 0.6 mm, which was deemed more than satisfactory for our system's construction. Opting for the 3x quality setting aimed to optimize scan quality while minimizing the scanning duration. These parameters were subsequently adjusted to maintain a comparable accuracy albeit with reduced precision, thus shortening the inspection time to less than 3 minutes. This modification catered to the demands of complex vehicular robot missions in practical scenarios. Furthermore, variations in color preferences were explored and tested, with the findings revealing that while color scanning did not influence accuracy, it did enhance the visual comprehension of the situations observed at inspection points.

Within the framework of this mission's objectives, a series of 3D laser scanning scenarios unfolded throughout the course of these experiments. The inspection protocol encompassed halts of the robotic platform at specific junctures, where subsequent 3D laser scan measurements were conducted. The primary aim of this validation process was to ascertain the seamless integration of the robotic system with the scanner interface, thereby enabling the execution of scans orchestrated entirely by the robotic mission.

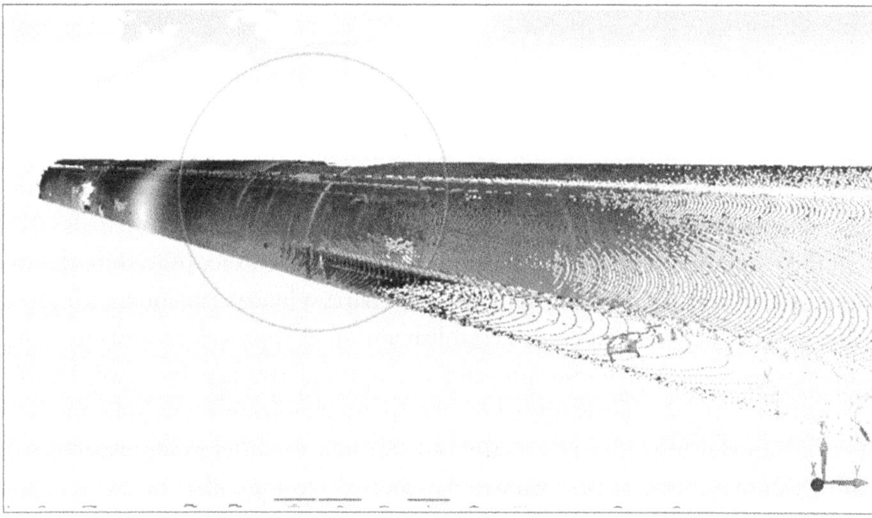

Figure 4.2. Collected point cloud at 2–3 mm accuracy.

Figure 4.3. Rendered image of a tunnel reconstructed in 3D.

Exemplary instances of localized inspections utilizing the laser scanner are illustrated below. Figure 4.2 showcases the amassed point cloud, characterized by millimeter-level accuracy. Meanwhile, Figure 4.3 offers a visual representation of the tunnel's 3D reconstruction. The collected data lends itself to diverse measurements and local computations, facilitated through various software applications and methodologies [3].

4.3.1 Equipment Capabilities and Setup

The initial dataset has played a pivotal role in acclimatizing the researchers to the forthcoming data measurements that will be undertaken and verified within the tunnel scenarios of the Egnatia Motorway. Additionally, this preliminary dataset has laid the groundwork for the autonomous robotic vehicle to execute laser scanning measurements automatically. The acquisition of this inaugural dataset has also facilitated the sequential tunnel measurements, mirroring the exact procedure that the robotic system will employ. Engaging with this dataset in experimental settings will further empower the researchers to establish a unified approach to measurements, encompassing accuracy, resolution, and the optimal strategy for amalgamating (co-registering) the distinct captured models (point clouds) into a singular model of a specific tunnel section, one of structural significance and interest for inspection. In the forthcoming months, an expanded dataset is slated for collection, with a more pronounced focus on precise inspection points within the tunnels (as indicative positions). This subsequent data collection phase will deliver heightened accuracy at points of direct inspection, thereby augmenting the overall efficacy of the endeavor.

The first validation of the approach took place from October 2020 to June 2021, with several visits to the site under investigation. For this, the above apparatus was used with the following inspection specifications:

- Point cloud number: 10.9M points.
- Point distance mm/10 m: 12.2 mm.
- Point distance mm/2.5 m: 3.1 mm.
- Inspection time (per scan): 5.5 minutes.
- Total inspection time (150 m of tunnel): 2.5 hours.
- Distance between consecutive scans: 30–50 m.

Validations and tunnel measurement continued during the project duration in 2022 and 2023 with tunnel deformation comparisons that proved the system accuracy to 1 mm that could be reduced further if higher resolution of the scanner was used or the scanner was positioned closer to the tunnel side under inspection. A visual representation of the test tunnel can be found in Figure 4.4 using the specifications above.

Other detailed models of the inspected areas can be seen in Figure 4.5, providing a full representation of a tunnel layout captured with 2 mm accuracy in measurements:

In the Figure 4.6 we present the concept of continuous measurements and possible escalation of any tunnel intrados deformation. This diagram does not represent actual tunnel measurements but is a hand made design to indicate the escalation and continuous monitoring concept. For purposes of confidentiality the actual measurements in the tunnels are not included in this publication.

Figure 4.4. Tunnel 3D reconstruction (external view).

Figure 4.5. Tunnel 3D reconstruction (external view).

4.3.2 Technical Approach

After the data collection at various positions in the tunnel with the laser scanner, the models had to be compared with each other to detect any actual deformation in the tunnel shape (and size). To accomplish that, approaches like ICP (Iterative Closest Point) where used as the first and most important aspect of the analysis included a proper (and correct/precise) co-registration of the different models (cloud points).

The ICP (Iterative Closest Point) in Figure 4.7, registration approach is a widely used technique in the field of computer vision, robotics, and 3D point cloud

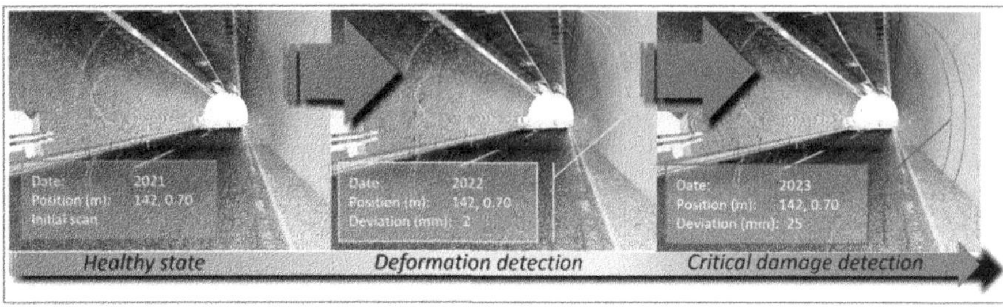

Figure 4.6. Concept of models comparison over time - deviation is totally experimental and does not correspond to actual measurements at the tunnel inspected. No deviation was found at the particular tunnel points. The diagram is only created for demonstration purposes of the escalation concept.

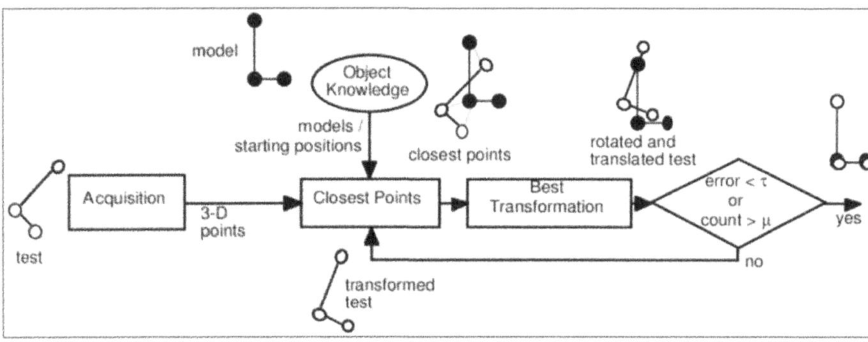

Figure 4.7. Iterative closest point algorithm.

processing. It is used for aligning and registering two or more sets of point clouds or 3D models to find their optimal spatial transformation, usually involving translation and rotation. The basic idea behind it, is to iteratively refine the transformation parameters until the best alignment between the source and target point clouds is achieved. ICP is effective when the initial transformation estimate is relatively close to the true alignment, and when there is sufficient overlap between the source and target point clouds. It's commonly used in scenarios like registering laser scans, aligning 3D models from different viewpoints, and aligning point clouds from sensors like LiDAR.

4.4 Work Novelty and Industrial Uptake

The innovation of the approach relies on the overall approach to combine different measurements over a tunnel or other structural element and create snapshots of its health status at particular points in time, while being able to detect deformations of its shape over time. This will further support the monitoring of escalation of any defect or other incident over the lifetime of the component. The innovation of the approach can be summarized below:

1. **Precision Measurements:** The foundation of the solution's novelty rests upon achieving unparalleled accuracy in measurements. This will ensure that every data point captured is reliable and contributes to a comprehensive understanding of the tunnel's condition.
2. **Expedited Tunnel Modeling:** The solution will revolutionize tunnel modeling by significantly expediting the process. This efficiency will be a hallmark of the product, streamlining the workflow and enabling faster decision-making.
3. **Human-Less Operation:** A key advancement lies in eliminating the need for human presence during inspections. The solution will function autonomously, reducing operational complexities and safety concerns.
4. **Automated Modeling and Aggregation:** Automation will be at the core of the solution's capabilities. The system will autonomously create models and aggregate data, minimizing manual intervention and potential errors.
5. **Seamless Application in Extended Tunnels:** The solution's impact will extend even to tunnels spanning over 3 kilometers in length. Its adaptability to such challenging contexts will underscore its versatility and effectiveness.
6. **Automated 3D Model Comparison and Defect Reporting:** An exceptional feature will be the solution's ability to automatically compare 3D models. It will swiftly identify discrepancies and generate reports for defect escalation. This process will enhance efficiency and facilitate prompt maintenance planning.

4.4.1 Extensions and Digital Assets

The approach validated above has proved the high value of precise deformation monitoring of motorway and other assets (civil structures, buildings etc.). It provides accurate and actual measurements of assets in the form of point-clouds and by comparing different versions of these we can obtain the deformation from one version to the next and thus monitor any defect escalation or other asset deformation.

This technology and approach has strong innovation potential to offer on its own as well as combined with other digital technologies such as Digital Twins and BIM (Building Information Modelling) approaches. 3D laser scanning plays a crucial role in creating accurate and dynamic digital twins that offer numerous benefits, ranging from improved asset management to enhanced safety and better decision-making in various industries. Some of the benefits for Digital Twinning as well as BIMs are summarized below:

1. **Accurate and Detailed Physical World/Asset Replication:** capturing of highly accurate and detailed data about the physical object and/or it surrounding environment supporting resembling the real-world parts.
2. **Time and Cost Efficiency:** significantly reducing time and effort required to capture data, making it a more cost-effective option for creating digital twins and BIM systems.
3. **Easier, Quicker and Real-time Updates:** models can be updated in real-time or at regular intervals. This allows for monitoring and tracking changes, damage, wear and tear, or any modifications made to the physical asset over time.
4. **Simulations, What-if and Analysis:** providing the data needed for simulations like structural analysis, energy consumption modeling, and fluid dynamics simulations, helping engineers and designers optimize their systems.
5. **Improved Maintenance, Safety and Operations:** support predictive maintenance including safety considerations by monitoring the condition of assets.
6. **Visualization and Situation Awareness:** support designing improvements or modifications to existing assets. Architects, engineers, and designers can visualize their plans in the context of the digital twin, ensuring compatibility and accuracy.
7. **Documentation, Reporting and Compliance:** easier and comprehensive documentation of assets or structures. Could be essential for regulatory compliance, preservation, and legal purposes.

Acknowledgements

This publication has been made following recent research results in the 'PILOTING', EC, co-funded, research project. This project has received funding from the European Union's Horizon 2020 research and innovation program (No 871542-5). The authors acknowledge the research outcomes of this publication belonging to the PILOTING consortium.

References

[1] The Importance of Road Maintenance, World Road Association, PIARC, 2014, Release 02.

[2] Communication From The Commission To The European Parliament, The Council, The European Economic And Social Committee And The Committee Of The Regions Towards A European Road Safety Area: Policy Orientations On Road Safety 2011–2020.

[3] K. Loupos, T. Krousarlis, P. Panetsos, E. Charalampakis – MOTORWAY TUNNELS RESILIENCE THROUGH AUTOMATED DEFORMATION MONITORING, XVI World Winter Service and Road Resilience Congress (PIARC 2022), 7–11 February 2022, Calgary, Canada, https://doi.org/10.5281/zenodo.6444151.

DOI: 10.1561/9781638282839.ch5

Chapter 5

Transforming Engineering Inspections Through Advanced UAVs

By Alejandro Murcia Crespo, Rafael Caballero González, Miguel Ángel Trujillo Soto, Francisco Javier Pérez Grau, Antidio Viguria, Ambar Garofano Soldado, José Ramiro Martínez de Dios and Aníbal Ollero Baturone

In the rapidly evolving landscape of engineering and technology, the quest for innovation continues to yield remarkable solutions that transform traditional practices and push the boundaries of what was once thought possible. The PILOTING H2020 European Project stands as a testament to the collaborative efforts and ingenuity of engineers and researchers who have united their expertise to address critical challenges in infrastructure inspection. This project marks a significant milestone in the realm of engineering inspections, introducing a fleet of cutting-edge robotic platforms that have the potential to revolutionize the way we examine and maintain critical structures.

The field of inspection and maintenance (I&M) encompasses a diverse array of methodologies aimed at ensuring the safety of infrastructure and personnel, all while effectively managing costs. These procedures involve meticulous examinations, often requiring elevation and the use of sensors in direct contact with the surfaces being scrutinized. Historically, these inspections have been carried out by technicians accessing specific inspection points using various methods such as man-lifts, cranes, scaffolds, or rope-access techniques. While inner inspections have been feasible for box-type bridges where individuals can access the interior, external assessments are also imperative.

In response to industry demand for innovative alternatives to manual inspections, the integration of sensors strategically positioned within bridges has been explored. These sensors provide localized measurements at a high frequency, ideal for closely monitoring critical structures on a small scale. However, this approach may not be optimal for conducting comprehensive inspections of larger bridge structures. The utilization of aerial robots, or drones, presents a promising solution to reduce costs, expedite inspection timelines, and enhance the overall quality and safety of the assessment process. Nevertheless, two primary technological challenges must be addressed: autonomous navigation of drones without relying on GNSS, enabling automated visual inspections, and the application of drones for contact inspections requiring detailed measurements on specific areas.

This chapter aligns with the goals and use cases of the PILOTING project, which centers on the adaptation, integration, and demonstration of robotic solutions within an integrated platform. This platform will undergo testing and evaluation across three major pilots: refineries within the Oil & Gas sector, bridges and viaducts in the Civil/Transport Infrastructure sector, and tunnels. Stakeholders across the entire value chain will participate in these pilots. The primary objectives of PILOTING include showcasing large-scale applications of robotics in Inspection and Maintenance (I&M), reducing commercial risks associated with robotics deployment, demonstrating the value and potential of robotics adoption, fostering an ecosystem around piloting I&M operations, and contributing to the establishment of industry standards for robotics in I&M.

To fulfill these objectives, PILOTING developed advanced robotic platforms tailored to the needs of the three industrial scenarios. The platform's deployment will underscore its tangible contributions to the inspection and maintenance community, highlighting its substantial socio-economic impact when applied on a larger scale. Through large-scale pilots conducted in authentic industrial environments, PILOTING will address key challenges in I&M by showcasing enhanced inspection and maintenance task rates, improved coverage, and performance, reduced operational costs and timelines, heightened inspection quality, and increased operator safety.

Among the remarkable achievements of the PILOTING project are the Unmanned Aerial Vehicles (UAVs) that have been meticulously developed to execute inspections in some of the most challenging and inaccessible environments. In this chapter, our focus converges on the UAV component of this endeavor, shedding light on four exceptional aerial platforms: AeroX, AERO-CAM, VIAD-DRONE, and TTDRONE. Each of these UAVs has been carefully engineered with a specific purpose, leveraging state-of-the-art technologies to address the unique demands of inspections in refineries, bridges, viaducts, and tunnels.

The AeroX UAV emerges as an epitome of precision engineering, embodying a groundbreaking approach to inspections through physical contact. By ingeniously integrating an aerial platform with a Robotic Mobile Contact Platform (RMCP), AeroX takes on the role of an aerial contact robot. This innovative design facilitates Ultrasonic Testing (UT) inspections, enabling meticulous examination of stationary surfaces that were once challenging to access. AeroX brings together the realms of aerial and tactile inspections, showcasing the project's commitment to pushing the boundaries of traditional methodologies.

AERO-CAM, on the other hand, redefines the concept of visual inspections for viaducts. Tailor-made to perfection, this Remotely Piloted Aircraft System (RPAS) boasts not only the quintessential characteristics of stability and robustness but also a deep understanding of the challenges posed by viaduct inspections. By embracing cutting-edge technology and a strategic configuration, AERO-CAM captures high-resolution images of viaduct underbellies, enabling engineers to analyze critical structural details with unprecedented clarity. This UAV operates beyond visual line of sight (BVLOS), epitomizing the project's aspiration to elevate inspection capabilities to new heights.

VIAD-DRONE underscores the significance of interaction in inspection tasks involving viaducts. Its specialized design facilitates physical interactions with viaduct structures, allowing tasks such as bearing checks, sensor box installations, and target setup to be executed efficiently and accurately. VIAD-DRONE encapsulates the ethos of the PILOTING project by offering a versatile solution for various inspection challenges linked by their reliance on physical interaction.

Lastly, in the subterranean realm, the TTDRONE takes center stage, fulfilling a pivotal role in tunnel inspections. Operating in tandem with an Unmanned Ground Vehicle (UGV) named CART, TTDRONE exemplifies seamless collaboration between airborne and ground-based platforms. This synergy empowers engineers to comprehensively assess tunnels, combining aerial perspectives with ground-level insights to create a holistic understanding of structural conditions.

As we unravel the technological advancements and autonomous capabilities of AEROX, AERO-CAM, VIAD-DRONE, and TTDRONE, we embark on a journey into a world where inspection becomes innovation, challenges find solutions, and engineering redefines the boundaries of possibility.

5.1 AEROX - Elevating Inspection to New Heights

The AeroX aerial contact robot stands as a pioneering UAV meticulously engineered for conducting meticulous physical contact inspections on stationary surfaces.

Its innovative design encompasses two distinct yet interconnected platforms: the aerial platform itself and the cutting-edge Robotic Mobile Contact Platform (RMCP), dedicated to Ultrasonic Testing (UT) inspections. This transformative system employs the RMCP, securely attached to the contact device, at the aerial platform's extremity.

Central to AeroX's revolutionary capabilities is its groundbreaking aerial robotic manipulator, a technological marvel that ushers in an era of unparalleled contact-based inspections. The manipulation mechanism seamlessly integrates a six-degree-of-freedom (DoF) robotic arm, a versatile robotic vehicle, and an adaptable end-effector furnished with inspection sensors and wheels. Operating within a semi-autonomous framework, AeroX's operational paradigm offers a multitude of advantages for contact-based inspections.

During the free-flight mode, the human pilot guides the robot until the end-effector establishes physical contact with the targeted surface. Upon contact initiation, AeroX seamlessly transitions into a fully autonomous, GNSS-free mode, orchestrating its position relative to the point of contact through the utilization of onboard sensors. While soaring autonomously, inspectors can adeptly manipulate the end-effector, ensuring continuous and precise contact with the inspection surface. This granular control permits the selection of specific points for inspections necessitating close sensor interaction.

The masterful control system embedded within AeroX effectively mitigates disturbances, seamlessly redirecting forces to the robot's center of mass. This strategic design empowers the aerial segments to absorb external forces, rendering the system resilient to external influences such as wind. Engineered with a robust 4-coaxial rotor configuration and a sleek yet sturdy design, AeroX guarantees unflinching stability, agile maneuverability, and unwavering resilience, even in the unlikely event of rotor malfunction. Its versatile architecture accommodates inspections across surfaces of varying orientations, seamlessly integrating into existing maintenance operations.

The journey of AeroX's development spanned multiple versions, with the second iteration serving as a foundation for validation and experimentation. Valuable insights gathered during this phase led to iterative improvements aimed at elevating the robot's overall Technology Readiness Level (TRL). The evolution to version 3 marked significant advancements such as:

- **Enhanced Structural Resilience:** The robot's structural integrity has been fortified, rendering it more resilient against unintended impacts. Moreover, the improved sturdiness of the robotic arm enhances the robot's precision during contact operations, attributed to its more accurate positional estimation.

- **Revised Design for Simplified Maintenance:** The cabling system has been reconfigured to ensure enhanced accessibility, facilitating prompt replacement in case of issues. Additionally, the overall structure has been streamlined for easier assembly, simplifying routine and unexpected maintenance tasks. This optimized design approach results in an aerial robot that exhibits superior modularity, enabling independent subcomponent replacements.
- **Streamlined Portability and Deployment:** In the last AeroX iteration, which required disassembly for transport by removing motor-connecting arms, the latest robot version features a foldable mechanism. This innovation enables effortless transportation and streamlined deployment upon removal from its container.
- **Integrated Sensors for Relative Surface Movements:** A relative localization sensor will be integrated with wheel movement encoders to generate a contextualized mapping of UT measurements, capturing relative movements across the surface.
- **Enhanced Dust and Light Rain Protection:** The dust and light rain protection mechanism has been upgraded to a more advanced system that offers easier installation and removal. Concurrently, this system serves as a safeguard for the robot's electronics.
- **Advanced Gas Detection System:** The previous gas detection system, which relied on a simple beeper for alerting gas detection, has undergone a transformation. The new system is compatible with a network of gas sensors and can be connected to ground-based or remote devices. This multipronged connectivity empowers diverse centralized alarm systems to receive gas-related alerts.

Notably, the manipulator arm in version 3 now communicates critical system information through an LED display, enhancing user interaction and facilitating real-time decision-making. The end-effector underwent substantial enhancements to enhance accuracy and minimize couplant wear, ensuring consistent and precise contact with the inspection surface.

The culmination of this rigorous developmental journey resulted in the final version of AeroX, meticulously engineered to adhere to initial specifications while incorporating all iterative enhancements (Figure 5.1). This unwavering commitment to refinement ensures AeroX's readiness for optimal performance, making it an invaluable asset for the ambitious endeavors of the PILOTING project.

AEROX, a pivotal component of the PILOTING project, demonstrates remarkable autonomous capabilities in the context of refinery tank inspections. Equipped with a specialized robotic arm for contact inspections, the AEROX flight platform is engineered to navigate and assess the interiors of refinery tanks with precision and efficiency.

Figure 5.1. Aerox version 3 aerial robot.

The refinery tank inspection scenario involves two distinct phases, each contributing to the platform's autonomous functionality.

During the approach phase, AEROX is guided by a pilot as it navigates from its take-off point to the contact zone on the tank's surface. The pilot's input is crucial at this stage, requiring careful maneuvering to establish initial contact with the tank. The platform's GPS receiver is utilized in this phase, aiding the pilot in ensuring accurate positioning. The location of the total station, which establishes a correlation between the platform's movement and a pre-created 3D map of the tank, is also accessible. The approach phase grants the pilot full control until the point of contact is achieved.

Once contact with the tank's surface is established, AEROX transitions into the autonomous contact inspection phase. This phase involves two crucial components: the manipulator arm and the Robotic Mobile Contact Platform (RMCP). During contact inspection, the GPS signal may encounter interference from metallic structures in the environment, leading to potential degradation. To address this, the platform relies on a network of onboard sensors situated along the manipulator arm and the RMCP. By performing inverse kinematics calculations through all degrees of freedom, AEROX computes its relative location with respect to the RMCP's position.

To achieve global localization, a critical step is taken involving a total station position estimator. This estimator leverages laser-tracked coordinates obtained from an onboard prism, which are then used to compute the RMCP's position in relation to the tank's surface (Figure 5.2). A numerical computation is executed to determine the precise position that adheres to the geometric constraints between AEROX and the tank. This transformation from relative to global localization ensures accurate positioning during the contact inspection phase, enhancing the platform's overall precision and autonomy.

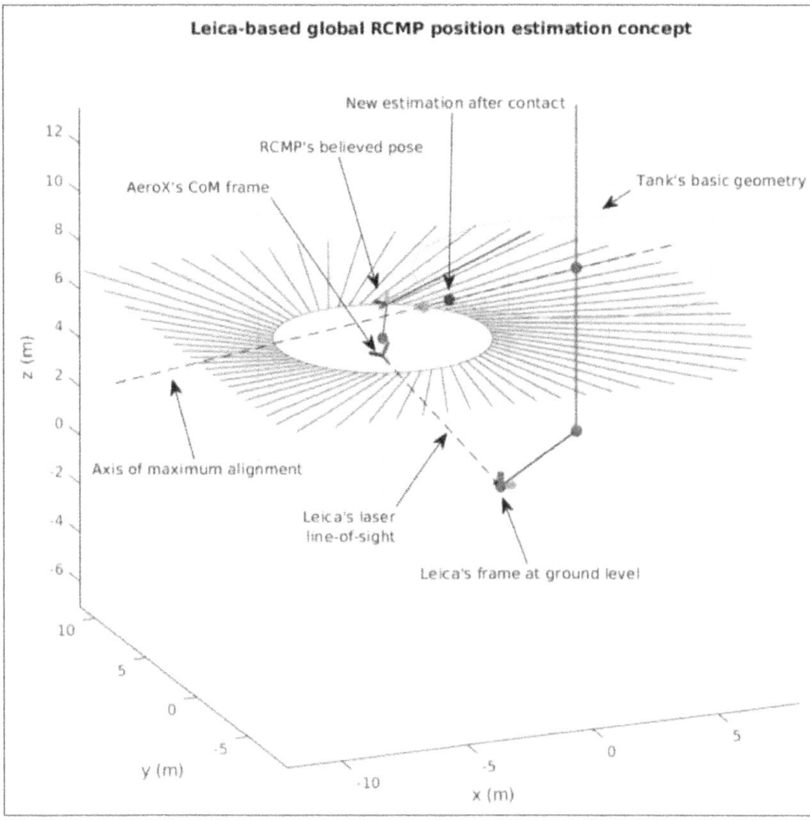

Figure 5.2. Position estimator algorithm visualization.

The validation process of AEROX's autonomous functionality is unique due to its reliance on the total station itself for the execution of inspections. As a result, comparisons with ground-truth data are not possible. Instead, the successful completion of the inspection serves as a validation of the localization algorithm. The millimeter-level accuracy ensured by the total station reinforces the confidence in the algorithm's effectiveness. The validation results, although unable to be directly compared to ground truth, contribute substantively to the project's overall success.

In summary, AEROX's autonomous capabilities within the context of refinery tank inspections showcase its ability to navigate, establish contact, and conduct accurate assessments within confined and complex environments. The integration of total station-based localization and sensor-driven positioning ensures precise and effective inspections, ultimately contributing to the project's mission of advancing inspection technologies in industrial settings.

In the context of oil refinery tank inspections, the AEROX platform's experimental results demonstrate its potential to revolutionize the efficiency and accuracy of ultrasonic thickness (UT) measurements. The need for extensive and precise inspections within oil refineries is underscored by the critical role that corrosion monitoring plays in maintaining safe operations. Corrosion-related equipment failures can lead to catastrophic accidents, making accurate wall thickness measurements essential.

In a standard-sized oil refinery, where 40,000 to 60,000 thickness measurement points are required within a 3-to-5-year interval, the AEROX platform presents a groundbreaking solution. The majority of inspection costs in such facilities (60% to 75%) are attributed to ultrasonic thickness measurements, which highlights the significance of this technology in maintaining refinery safety.

AEROX's experimental deployment at Chevron's Oronite plant showcased its ability to perform UT measurements on a large storage tank, eliminating the need for workers to operate at heights and significantly reducing access time and costs. The chosen experiment area in the uppermost left corner of the tank farm allowed for thorough validation of AEROX's capabilities.

The experiments involved two primary profiles: a longitudinal ultrasonic scan along the vertical axis, covering most of the tank's height, and a predefined path encompassing an area surrounding numerical markings. The validation process compared AEROX's UT measurements against manual readings using a hand tool. Initial results indicated some inconsistencies, with a variation of approximately 0.3 to 0.5 mm error in accuracy compared to manual readings. Factors contributing to these discrepancies included signal post-processing issues, noise, and the alignment of the probe's sensor with the inspected surface.

Despite these initial challenges, AEROX's potential was evident when using a clean calibration block in a controlled environment. The platform's UT probe was capable of providing accurate readings, pointing toward the need for improvements in signal post-processing algorithms and mechanical design to enhance accuracy and reliability during dynamic operations.

During the experiments, an advanced contact control algorithm enabled AEROX to maintain contact with the tank's surface while flying autonomously (see Figure 5.3). This automation significantly streamlined the inspection process, allowing the operator to focus solely on positioning the UT sensor and taking measurements.

The integration of the total station for high-precision positioning, coupled with the advanced control algorithm, facilitated spatial correlation of UT measures with specific 3D points on the tank. This repeatability ensures that inspections can be performed at the same location in different sessions, providing valuable consistency for end-users.

Overall, AEROX's experimental results showcase its potential to transform refinery tank inspections by reducing human involvement, minimizing costs, and

Figure 5.3. AeroX platform performing a line scan.

enhancing safety through accurate and autonomous UT measurements. Ongoing efforts to refine signal processing algorithms and improve mechanical design promise even greater precision and reliability in future deployments.

5.2 AERO-CAM - Clear Visions From Above

The AERO-CAM Remotely Piloted Aircraft System (RPAS), a tailor-made DJI Matrice 600 Pro, stands as the ultimate aerial platform, meticulously crafted to facilitate visual inspections of viaducts (Figure 5.4). This hexacopter UAV boasts an unwaveringly robust configuration, featuring a sextet of powerful motors that not

Figure 5.4. AERO-CAM aerial robot.

only ensure optimal performance but also provide an inherent layer of resilience through redundancy, mitigating any potential fallout from engine failures. At its core, this engineering marvel strikes a delicate balance between compactness and endurance, enabling extended flight times to accommodate critical payloads such as the high-resolution camera and the indispensable localization sensors. Operating beyond the confines of visual line of sight (BVLOS), the camera occupies a vantage point atop the UAV, positioned strategically to capture comprehensive images of the viaduct's underbelly, thereby affording unparalleled clarity when examining horizontal surfaces.

Anticipating the inevitable GNSS signal attenuation beneath the viaduct's imposing structure, the AERO-CAM seamlessly incorporates a state-of-the-art LIDAR system. This integration not only enriches the system's situational awareness but also confers an unprecedented level of reliable localization. The symbiotic marriage of hardware and technology encompasses a meticulously calibrated gimbal-mounted Sony alpha 7R IV camera, meticulously fine-tuned for the rigors of inspection missions. This optical prowess is further augmented by the inclusion of a Gremsy T3 gimbal stabilizer, which, through a bespoke damper system, systematically curtails vibrations during flight, ensuring pristine imaging quality even in turbulent conditions. As the backbone of communication, the AERO-CAM relies upon a dual-channel Wi-Fi antenna arrangement, orchestrating seamless connections between the camera PC and the aircraft's companion computer, ultimately extending its reach to the realm of ground stations.

While the foundational hardware architecture remains resolute, the true evolution of the AERO-CAM resides in its software enhancements, meticulously designed to catapult the system into the realm of advanced autonomous functionalities, with a singular focus on localization precision and flight control algorithms. The finalized AERO-CAM system configuration coalesces around several key components, each contributing to the system's exceptional capabilities:

- **Airframe:** The DJI Matrice 600 Pro, seamlessly integrated with the A3 Pro autopilot, Lightbridge 2 HD communication system, and the intelligence of Smart Batteries.
- **Camera:** The Sony alpha 7R IV, meticulously calibrated and optimized to excel in the realm of inspection missions, capturing intricate details with unmatched precision.
- **Gimbal and Damper System:** The Gremsy T3 gimbal, fortified with a bespoke damper system, tirelessly striving to maintain unwavering stability while minimizing vibrations, thereby ensuring the delivery of immaculate imagery.
- **LIDAR:** The Ouster LIDAR sensor (model OS0 with an impressive 128 channels) stands as the cornerstone of reliable positioning estimation, rendering the AERO-CAM impervious to the signal challenges often posed by the viaduct's structure.
- **Communication System:** A symphony of communication channels, orchestrated by both the on-board camera PC and the Ubiquity module, synergize to foster seamless interactions between companion computers and ground stations, culminating in a harmonious flow of information.

In summary, the AERO-CAM Unmanned Aerial System (UAS) emerges as an epitome of efficiency and reliability, a versatile solution meticulously designed to cater to the multifaceted demands of visual inspection endeavors upon viaducts. Armed with an impressive array of advanced components and functionalities, this aerial marvel emerges as the quintessential tool for conducting accurate, comprehensive, and insightful assessments, ushering in a new era of precision-driven inspection practices.

AERO-CAM's autonomous functionalities for viaduct visual inspection involve a sophisticated system that ensures accurate localization and positioning. This high-resolution robotic platform, equipped with a camera mounted on a gimbal, employs a combination of sensors, including LIDAR, IMU, and altimeter, all managed by an Intel Nuc i7 onboard computer.

The localization process consists of two algorithms: global localization and relative localization. For global localization, the Direct LIDAR Localization (DLL) algorithm is employed (Figure 5.5). This approach relies on a pre-constructed 3D map created with a total station, representing a point cloud that establishes the coordinate system. The algorithm performs trilinear interpolation using the

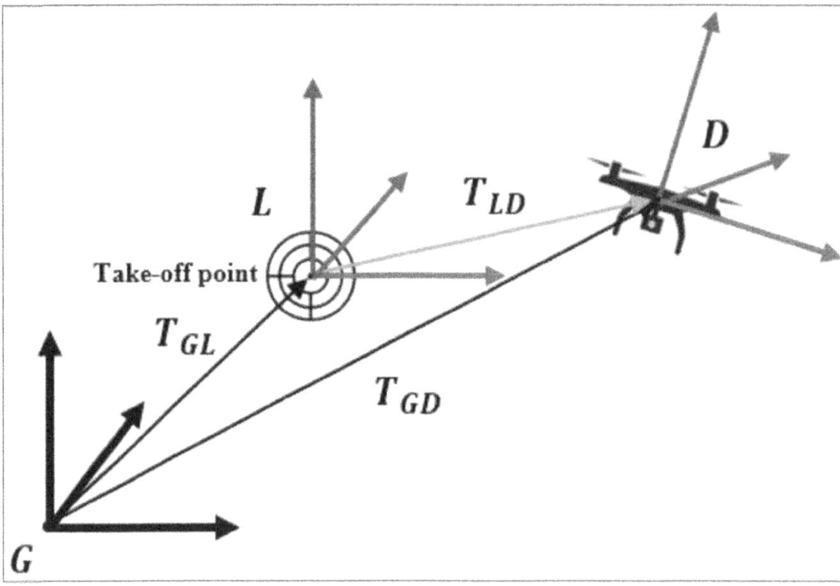

Figure 5.5. AERO-CAM full transformation system.

pre-computed map and the incoming LIDAR point cloud. The quality of the initial guess significantly affects algorithm performance, with experimental results demonstrating optimal ranges for accurate localization. This global localization algorithm is executed on the ground computer before takeoff, ensuring consistent coordination between the UAV and the map during inspections.

The relative localization algorithm, executed on the onboard computer, utilizes LIO-SAM, a tightly coupled fusion of LIDAR and IMU data. It calculates the UAV's pose with high frequency, supplying essential data for flight control and mission completion. Enhancements have been made to improve algorithm performance, addressing issues such as asynchronous IMU messages and gravity calibration.

Validation of the localization algorithms has been conducted on a real viaduct, comparing results to ground truth measurements obtained with a total station. The improved implementation of the localization algorithm demonstrates enhanced accuracy and smoothness, validating AERO-CAM's ability to perform precise and reliable inspections.

In conclusion, AERO-CAM's autonomous functionalities for viaduct inspection encompass a comprehensive (Figure 5.6) system that integrates advanced algorithms, sensors, and computational processes to ensure accurate localization and optimal performance during visual inspections.

The viaduct visual inspection experiments employing AERO-CAM revealed the efficacy and precision of the autonomous functionalities embedded within the

Figure 5.6. 3D visualization of the localization algorithm without GNSS.

robotic platform. These experiments encompassed three crucial phases: general visual inspection, surveying target installation, and specific visual inspection.

5.2.1 General Visual Inspection

During the general visual inspection phase, AERO-CAM autonomously executed a series of flights to thoroughly assess the viaduct's condition. The platform utilized its high-resolution camera, integrated with artificial intelligence, to capture images of the infrastructure and detect potential defects.

One of the key accomplishments was the remarkable repeatability demonstrated by the robot. Trajectories from different flights were superimposed onto a 3D model, showcasing consistent inspection paths and locations. This repeatability is pivotal for maintaining inspection accuracy across multiple missions.

Importantly, the entire inspection process, from takeoff to landing, was carried out autonomously. The robot's robust LIDAR-based localization system ensured accurate positioning, even in GNSS-denied environments. Additionally, an obstacle detection system bolstered mission safety, particularly in situations where visual line of sight was obstructed.

Real-time monitoring of the inspection missions was facilitated through the ground control station (gRCS) and mobile control station (mRCS). These stations provided comprehensive data visualization and status updates, ensuring effective mission oversight.

5.2.2 Surveying Target Installation

In the surveying target installation phase, the VIAD-DRONE robot was deployed to install surveying targets near identified inspection points. The robot autonomously navigated to the designated areas and made physical contact with the viaduct's surface to affix the millimeter-patterned targets.

Notably, the robot's localization accuracy, estimated autonomously without GNSS, was validated against reference data obtained from a Leica Total Station. The results highlighted the system's robustness and reliability, reinforcing its capability for accurate pose estimation in challenging scenarios.

The successful installation of surveying targets near the same inspection location on different days and under varying wind conditions underscored the repeatability and versatility of the platform.

5.2.3 Specific Visual Inspection

With surveying targets in place, AERO-CAM executed specific visual inspections to capture detailed images around the targets. Multiple attempts were made to photograph the same location, and despite minor deviations caused by flight fluctuations, the platform consistently achieved its imaging objectives.

Collectively, these experimental results showcase the successful integration of advanced robotics, sensor technologies, precise localization algorithms, and AI-powered image analysis. The AERO-CAM platform demonstrated its potential to significantly enhance the efficiency, accuracy, and safety of viaduct visual inspections. The combination of autonomous flight, accurate pose estimation, and repeatable mission execution holds promise for revolutionizing infrastructure assessment practices.

5.3 VIAD-DRONE – Precision Interaction

The VIAD-DRONE is a specialized flying platform created for various inspection tasks involving viaducts. In the PILOTING project, this platform is intended to perform tasks like checking bearings, installing sensor boxes, and setting up targets. All these tasks share a common factor: the flying robot physically interacts with the viaduct structure. This sometimes requires accessing tight spaces, such as the area between the top of a pillar and the deck. To address this, the platform needs to be compact enough to navigate these spaces, while also being strong enough to carry the necessary tools for working on the viaduct. The proposed solution involves a modular design. The core platform can be customized with different attachments

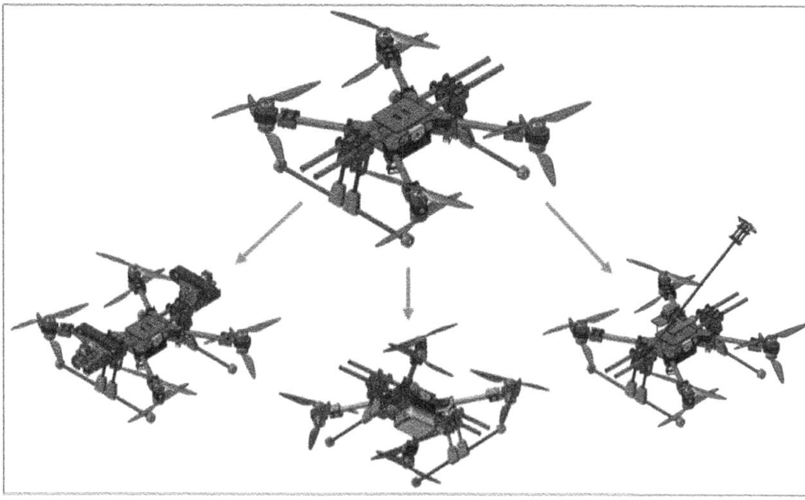

Figure 5.7. VIAD-DRONE modular concept.

to suit the specific requirements of each task. In Figure 5.7, you can see the modular concept with the three main attachments, each tailored for one of the use cases.

The base platform, common for every configuration, includes the airframe with the electronics and batteries, the onboard computer, the communications system, and the navigation sensors.

The VIAD-DRONE aerial robot was purposefully created to operate effectively in very limited spaces, including direct interaction with viaduct structures. This introduces significant constraints in terms of size and the weight it can carry, which in turn limits the sensors and processing hardware that can be installed on the robot.

In contrast to using a single, high-performance sensor as the core of the perception system (as initially planned), a different approach was taken during the second phase of development. The perception system for VIAD-DRONE was redesigned to rely on combining measurements from various types of sensors, a strategy known as sensor fusion. This approach offers several benefits:

1. **Weight Reduction:** Instead of relying on a single heavy sensor, the new system integrates several lightweight sensors. This helps decrease the overall weight of the sensors carried on the robot.
2. **Lower Computational Demand:** Processing measurements from a single high-performance sensor requires significant computational power. In contrast, extracting information from each type of sensor for the robot's pose estimation demands less computational resources.
3. **Enhanced Robustness:** By obtaining pose estimation from a range of different sensors rather than relying solely on one, the system gains greater resilience and reliability.

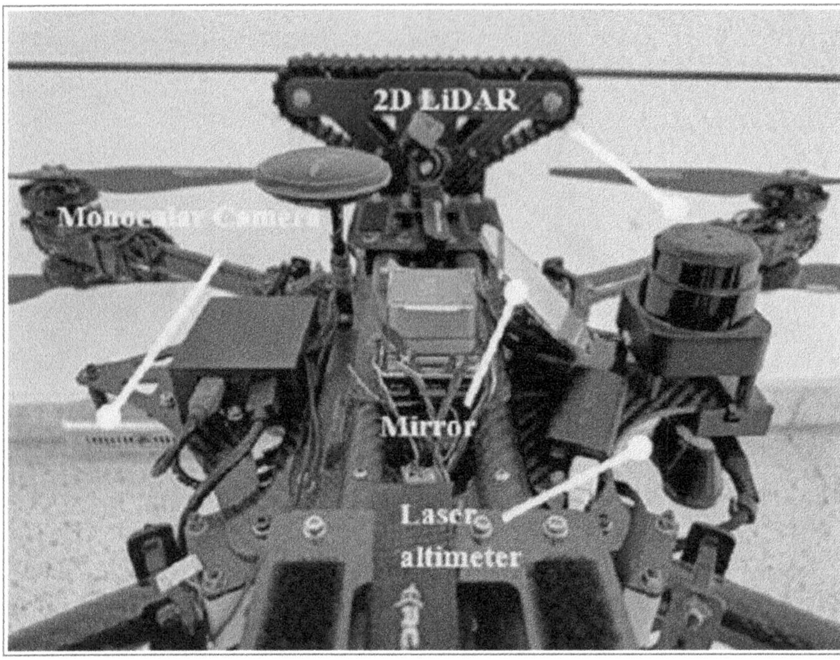

Figure 5.8. VIAD-DRONE sensors setup.

These adjustments not only address the constraints imposed by the robot's size and payload capacity but also enhance its overall performance in navigating and interacting with viaduct structures.

The VIAD-DRONE is a specialized aerial robot designed to carry out various inspection tasks on viaducts. These tasks include inspecting viaduct bearings, installing sensors, and placing targets (Figure 5.8). The robot physically interacts with the viaduct structure, often maneuvering through tight spaces between pillars and the deck. To meet these demands, the VIAD-DRONE is engineered to be compact yet powerful, equipped to carry the necessary tools.

In the case of bearings inspection, the robot employs a camera to capture images of the viaduct bearings. The camera is equipped with stabilization technology to ensure high-quality images. For sensor installation tasks, the robot autonomously positions a sensor box inside a designated hole on the viaduct pillar. This box collects essential structural measurements like vibrations. Both of these tasks involve the robot making direct contact with the viaduct deck and utilizing caterpillar-like mechanisms to navigate.

In missions focused on target installation, the VIAD-DRONE deploys markers onto specific locations of the pillar. These markers are utilized to identify any defects or issues in the structure. These missions encompass a series of steps, including take-off, navigation, and establishing physical contact with the viaduct structure.

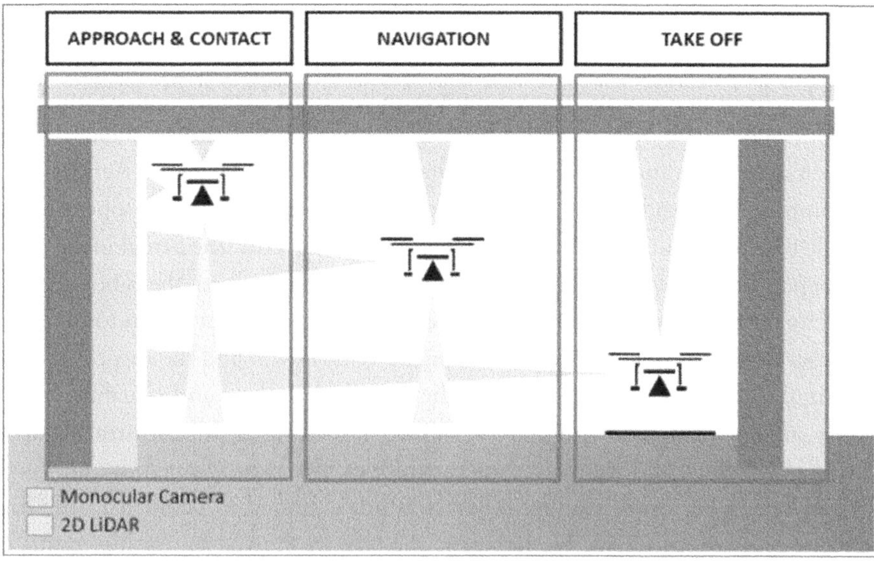

Figure 5.9. Stages of the viaduct inspection missions: take off, navigation, and approach & contact.

The localization method of the VIAD-DRONE, which determines its precise position, relies on a fusion of data from different sensors. This includes a camera, a laser altimeter, and a LiDAR (Light Detection and Ranging) system (Figure 5.9). The robot's movement is estimated by integrating visual odometry data from the camera and altitude measurements from the altimeter. The 2D LiDAR system is crucial for locating fixed objects like viaduct pillars and decks, enhancing the accuracy of positioning.

To effectively combine the outputs of these sensors for accurate localization, the robot employs a Kalman Filter. This filter aids in predicting the robot's position and subsequently updating it based on the information from the sensors. Real-world experiments conducted in viaduct scenarios demonstrated the success of this method in providing reliable and precise robot localization. The average errors in different axes were exceptionally low, affirming the efficacy of the system.

In essence, the VIAD-DRONE adeptly carries out viaduct inspection tasks by leveraging a range of sensors and advanced algorithms to achieve accurate and dependable positioning.

The VIAD-DRONE project conducted significant experiments involving viaduct inspection and sensor installation. One major aspect was the inspection of viaduct bearings. These bearings are critical to a viaduct's structural integrity, and their degradation can lead to defects. Visual inspections are traditionally carried out by human inspectors, but the VIAD-DRONE project aimed to replace this with aerial robots. The first pilot experiments focused on the visual inspection of bearings on the Álora viaduct. The VIAD-DRONE was designed to approach the viaduct, attach to the deck, and capture close-up images of the bearings. The robot's localization relied on various sensors, achieving a high level of accuracy.

Another key experiment involved sensor installation on viaduct pillars. These sensors monitor vibrations and accelerations in the viaduct's structure. Aerial robots were proposed to install these sensors, avoiding the need for human operators in difficult-to-reach locations. However, due to safety concerns, the actual installation was not performed on the viaduct pillars in the first pilot. Instead, the robot demonstrated its manipulation capabilities on the ground. The project aimed to integrate the sensor data with the Digital Monitoring System (DMS) and the Inspection and Visualization Platform (IVP) for comprehensive data analysis and visualization.

The project demonstrated that the VIAD-DRONE could perform accurate bearing inspections and showcased its capabilities in sensor installation, although the installation on top of viaduct pillars was not executed as planned in the first pilot due to safety concerns. Nonetheless, the project showed promising results in enhancing inspection efficiency, reducing costs, and increasing safety through the use of aerial robots and advanced sensors.

5.4 TTDRONE – Navigating the Subsurface with Expertise

The TTDRONE is an integral component of a specialized platform designed for tunnel inspection purposes. This platform consists of two key elements: an Unmanned Ground Vehicle (UGV) referred to as CART, outlined in a specified section, and an Unmanned Aerial Vehicle (UAV), the subject of detailed description in the current section. The platform's composition is visually depicted in Figure 5.10, aiming to comprehensively inspect tunnel structures through a combination of data acquisition and analysis.

At its core, the TTDRONE serves as an aerial robot tailored for tunnel inspection tasks. Its primary function is to conduct localized assessments within the tunnel environment, complementing the broader inspection activities conducted by the UGV, CART. By capturing close-up images and precise measurements, the TTDRONE is capable of examining inspection points that are inaccessible to the ground-based robot.

A distinctive feature of the TTDRONE is its connection to the UGV via a tethered system. This arrangement facilitates the transmission of both power and data between the UGV and the UAV through a cable connection. The term "TTDRONE" itself is an abbreviation for "Tunnel Tethered DRONE," which succinctly describes its role and mode of operation.

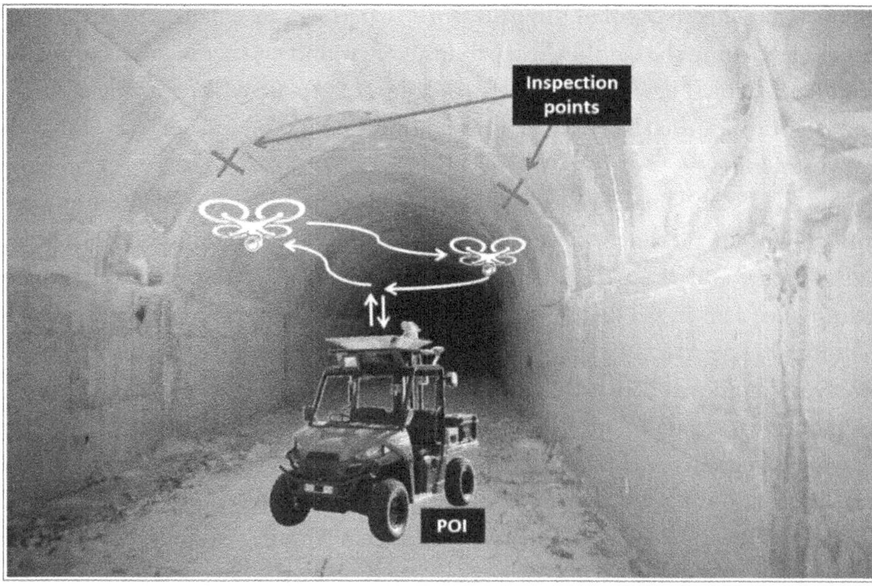

Figure 5.10. Tunnel inspection robotic system integrates a UGV and a UAV.

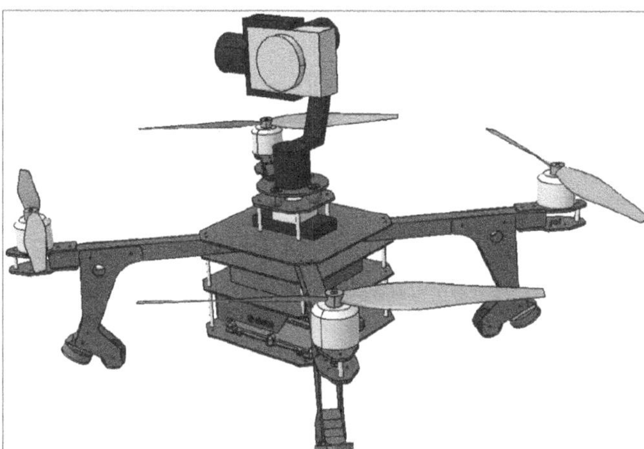

Figure 5.11. CAD of the TTDRONE.

Figure 5.11 provides a visual representation of the initial CAD design of the TTDRONE. Although a physical prototype of the TTDRONE has not been constructed at the time of documenting, this circumstance is attributed to international procurement delays. Notwithstanding this setback, preparations are underway to finalize the TTDRONE for integration and preliminary field experiments anticipated in the upcoming year.

Each constituent component of this comprehensive platform, including the UGV (CART) and the UAV (TTDRONE), is expounded upon in the subsequent section. The collective goal of this platform is to perform an exhaustive inspection of tunnels, leveraging the combined capabilities of ground and aerial robots to acquire images and data. These datasets are then subjected to analysis by artificial intelligence algorithms, thereby enabling the identification of potential defects within the tunnel structure.

The TTDRONE aerial robot is specifically designed to function autonomously within constrained tunnel environments, which impose limitations on its size, payload capacity, and the sensors it can carry. Instead of relying on a single high-performance sensor, TTDRONE employs a multi-modal perception system that incorporates several sensors. This approach offers benefits such as reduced weight, lower computational demands, and increased robustness.

One critical capability of TTDRONE is its ability to localize itself in relation to the Unmanned Ground Vehicle (UGV). While the UGV possesses its own localization system and can self-localize globally within the tunnel's coordinate system, the TTDRONE needs to establish relative localization with respect to its take-off position on the UGV's roof. This relative pose can then be transformed into the global tunnel coordinate system using the UGV's global localization data. This process ensures that the defect images captured by TTDRONE are referenced correctly in the tunnel's global coordinates.

To achieve this localization, TTDRONE is equipped with a sensor setup detailed in Figure 5.12. The sensor configuration has undergone iterative redesign to enhance robustness and reduce uncertainty. The final sensor setup comprises the following components:

1. **2D LiDAR (LiDAR1):** Positioned transversely to the tunnel's main axis, this LiDAR measures and processes data to determine the robot's localization in the plane perpendicular to the tunnel's main axis.

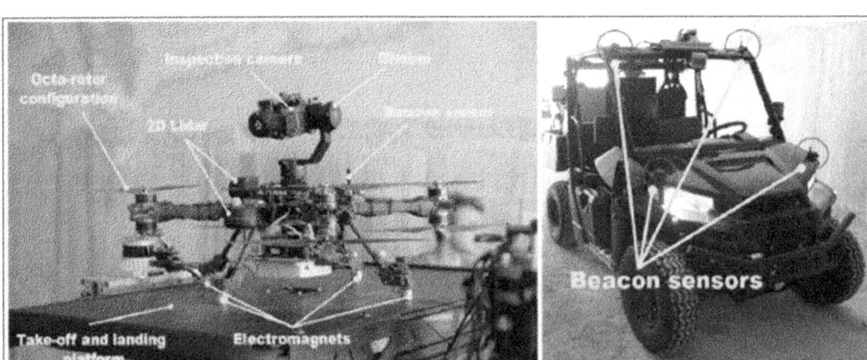

Figure 5.12. (Left) Setup of TTDRONE in tunnel inspection. (Right) Setup of the four UWB beacons on UGV used for TTDRONE localization.

2. **2D LiDAR with Horizontal Plane (LiDAR2):** This LiDAR is oriented with a horizontal sensing plane. It provides measurements used to calculate the aerial robot's yaw angle relative to the direction of the tunnel's main axis.
3. **UWB Beacon:** The TTDRONE incorporates a UWB beacon, which estimates the robot's location in relation to four UWB beacons situated on the UGV. This UWB-based localization aids in establishing accurate relative positioning.

The overarching objective is for TTDRONE to autonomously compute and maintain its pose estimations concerning its take-off and landing positions on the UGV's roof. Through this advanced sensor setup and localization approach, TTDRONE is empowered to perform effective defect inspection and data collection in challenging tunnel environments.

The TTDRONE underwent a series of comprehensive experiments conducted in collaboration with Egnatia Odos AE (EOAE), representing tunnel inspection operations on actual segments of the transport network. These experiments spanned a duration of 9 days and involved the careful orchestration of EOAE, INLECOM, and other partners. The experiments were organized in two different tunnels to serve as deployment and integration exercises, facilitated by EOAE's provision of infrastructures, safety measures, and other logistical support.

5.4.1 General Tunnel Inspection

- Experiments were conducted in the Metsovo tunnel and a nearby drainage tunnel. Metsovo tunnel was chosen for its interesting inspection points within live traffic scenarios, while the drainage tunnel provided a controlled environment to simulate tunnel construction.
- Various scenarios were designed to cover the wide array of robotic missions, testing procedures, and configurations under realistic conditions, including traffic and weather.
- Key scenarios focused on validating the capabilities and functions of the robot, autonomous navigation, and inspection missions.
- Robot capabilities and function verification included measurements of dimensions, weight, clearance, suspension functionality, battery status, emergency shutdown, and control override.
- Autonomous navigation was tested in a tunnel without GNSS coverage and fixed lighting. Landmarks were used for relocalization and drift correction.
- Inspection missions involved the robot autonomously navigating through waypoints, capturing images of points of interest, controlling the pan and tilt mechanisms for image capture, and performing 3D laser scans.

5.4.2 Local Tunnel Inspection

- In this scenario, both ground and aerial robots were utilized for more thorough inspections. The aerial robot inspected difficult-to-reach areas detected by the ground robot.
- Inspection plans were carefully designed based on expert analysis of tunnel defects, incorporating different types of defects and considering tunnel characteristics.
- The aerial robot autonomously navigated through specified waypoints, capturing high-quality images of defects using precise gimbal control.
- 3D laser scanning was also employed in local inspections, providing millimeter-accurate 3D reconstructions of tunnel areas with defects.
- The inspection results were uploaded to a central platform for analysis and assessment, aiding defect identification and performance evaluation of the robotic system.

Scenario 1. Aerial Robot Inspection Mission:

- The aerial robot conducted visual inspections of the drainage tunnel, capturing images of previously detected defects using image stabilization devices.
- The inspection plans were designed based on accurate maps, with inspection points precisely located.

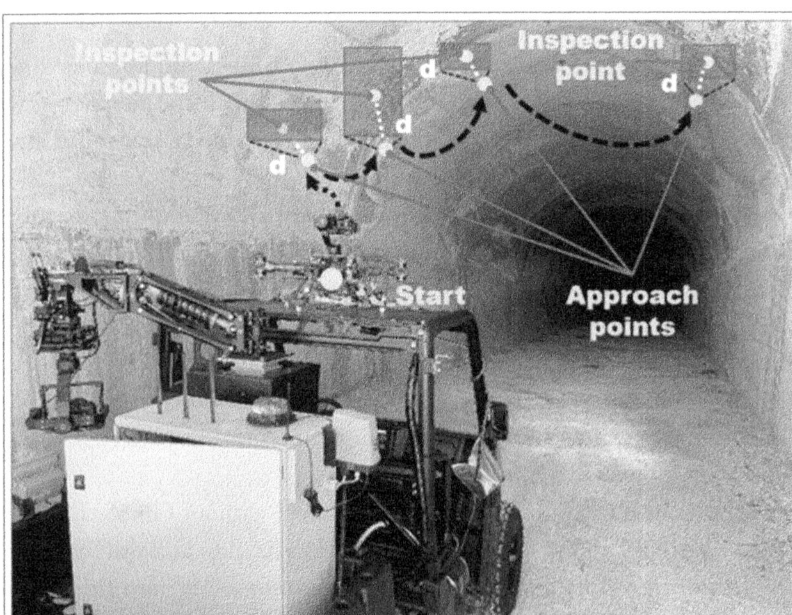

Figure 5.13. Tunnel local inspection procedure developed by the TTDRONE.

Figure 5.14. TTDRONE hardware setup used (left) and the 3D printed structure built (right) for the Scenario 1 experiments.

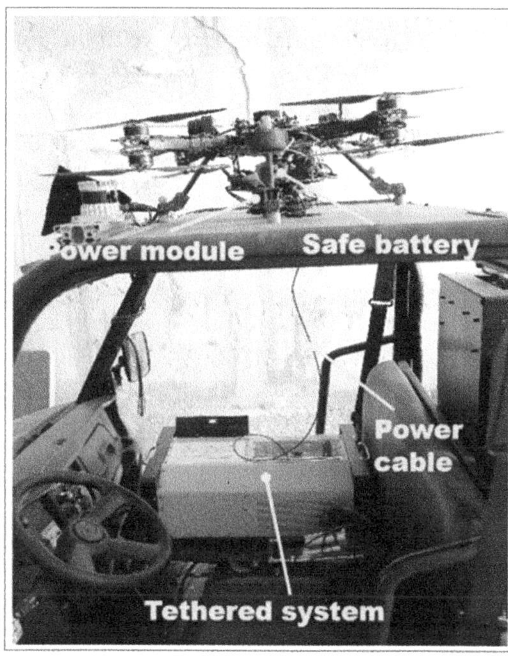

Figure 5.15. TTDRONE hardware setup for the Scenario 2 experiments. Also, the tethered system is shown.

- The aerial robot's flight trajectory, gimbal orientation, and image capture were controlled autonomously using the ground control station.
- AI-based defect identification was utilized to identify defects in captured images, with preliminary results showing promise in defect detection.

Scenario 2. Tethering Tests:

- In this scenario, the aerial robot was tethered to the ground robot, aiming to demonstrate reliable tethered flights in a real tunnel environment (Figure 5.15).

- The cable passed from the aerial robot through the roof of the ground robot to a power module.
- The tethered flights were focused on validating the tethering system's functionality and reliability.
- The aerial robot's hardware configuration differed from Scenario 1, with the inspection camera and other components removed for safety and payload reduction.

In both scenarios, these experiments aimed to validate the system's functionality, performance, and robustness in real tunnel environments. The results showcased successful navigation, defect identification, and data collection capabilities of the TTDRONE, indicating its potential for effective tunnel inspection and defect detection applications.

DOI: 10.1561/9781638282839.ch6

Chapter 6

Visualizing the End-to-End Workflow of Digital Inspections Through the Intelligence and Visualization Portal

By Marianna Corinaldesi, Arnaud Andrianavalomahefa, Safidy Ratsimbazafy, Patrick Ursolino, Elena Chiggiato and Luc Stakenborg

The increasing complexity of modern Inspection and Maintenance (I&M) activities, especially in the context of ageing infrastructure, demands novel digital solutions. Traditional methods, often paper-driven and manual, are becoming obsolete in the face of evolving challenges such as intricate asset structures, increased regulatory scrutiny, and the necessity for precision and repeatability. Funded by the European Union, the PILOTING project introduces a comprehensive digital approach to I&M, integrating multiple robotic systems for enhanced efficiency.

In this chapter, we present the Intelligence and Visualization Portal (IVP), a platform meticulously designed for the digital management of asset integrity and inspection workflows. In contrast to conventional systems, the design of the IVP is not confined to any specific robots, ensuring its compatibility and adaptability across a wide range of robotic systems. This ensures adaptability and scalability, making it compatible with an array of robotic systems, from ground-based units to aerial drones.

The pivotal feature of the IVP is its emphasis on 3D digital twins – precise digital replicas of real-world assets. These digital counterparts serve multiple functions. They act as foundational blueprints during inspection planning, enabling users to accurately pinpoint specific areas to inspect, as well as defining the tasks to perform during the inspection processes. Digital twins also provide robots with a spatial reference model, ensuring that every inspection is consistent and repeatable. Post-inspection, the digital twins become interactive canvases, where collected data is visualized, analyzed, and contextualized, transforming simple data into actionable insights.

Integration is a recurring theme in IVP. It interfaces with the PILOTING IoT platform, as well as several AI systems to augment its capabilities by automating data analytics. This synergy facilitates not only the collection of data but also its interpretation. Advanced AI algorithms automatically identify, classify, and annotate potential defects, expediting the analysis process and minimizing human errors.

Field experiments in diverse environments, such as refineries, viaducts, and tunnels, have demonstrated the IVP robustness and versatility. These real-world tests have highlighted its capability to handle complex scenarios, vast datasets, and diverse asset structures.

In conclusion, the Intelligence and Visualization Portal, introduced by the PILOTING project, clearly showcases the revolutionary impact of digital solutions in the field of inspection and maintenance. It encapsulates the future of I&M – a future that is digital, integrated, precise, and above all, user-centric.

6.1 The PILOTING Platform

European infrastructures, particularly refineries and civil constructions, are displaying signs of age and deterioration. As these vital assets age, the urgency to inspect and maintain them intensifies. However, the traditional ways of conducting these inspections, often ad-hoc, are becoming less proficient and increasingly risky. There is a pressing need to enhance the quality and effectiveness of I&M activities, not just to maintain the functionality and safety of these structures, but also to ensure the safety of the workers and to optimize the costs. This situation necessitates a transformation in how inspections are carried out.

The PILOTING Platform (see Figure 6.1) was conceptualized to address this complex challenge. The project's core strategy revolves around the adaptation, integration, and demonstration of robotic solutions. These robotic solutions are deployed in large-scale pilots and supported by an integrated platform designed specifically to digitalize and enhance the entire inspection workflow.

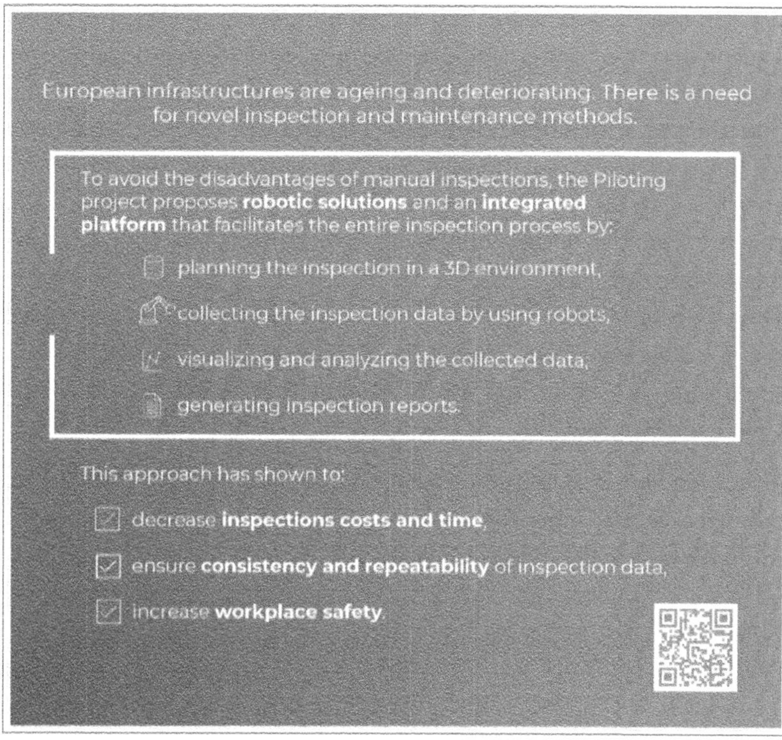

Figure 6.1. Overview of the PILOTING project (https://piloting-project.eu/).

The benefits of the PILOTING system are manifold:

- **Efficiency in Inspections:** It is designed with a primary focus on productivity, also resulting in notable reductions in both the time and financial resources traditionally associated with inspections.
- **Reliable Data Collection:** Coupled with this competency is the platform's ability to gather comprehensive and accurate inspection data, ensuring that any information used for decision-making is reliable.
- **Safety Enhancements:** Safety, a paramount concern in inspection activities, is significantly enhanced through the PILOTING approach. By leveraging robotic solutions, the inherent risks of manual inspections, especially in potentially hazardous environments, are substantially mitigated. This ensures not only the safety of the infrastructure but also the well-being of the inspection personnel.
- **Consistency Over Time:** Beyond these immediate advantages, the PILOTING system ensures consistency and repeatability of inspections, while guaranteeing that standards and procedures are respected. Indeed, with the integration of robots and the PILOTING system capabilities, inspections

become not just repeatable but consistently accurate. This is invaluable for asset owners who need localized inspection data that can be reliably compared over time, ensuring that every potential issue is promptly addressed.

- **Streamlined Inspection Planning:** The platform is more than just a tool for data collection; it streamlines the entire inspection planning process. It adeptly organizes all pertinent data, from the insights gleaned during inspections to essential documents like asset blueprints, safety regulations, and required permissions.
- **Data-Driven Decision Making:** This cohesive organization, combined with the rich data repository, empowers decision-makers to adopt a more data-driven approach, ensuring that any maintenance or corrective actions are both timely and effective.

In the subsequent sections, we will delve deeper into the PILOTING Platform, exploring its various components and understanding how it harmoniously integrates to revolutionize the inspection landscape.

6.2 The Intelligence and Visualization Portal

In a rapidly modernizing world, the ways we interact with, maintain, and manage assets have undergone a transformative shift. The traditional methodologies that dominated the inspection landscape are now giving way to more innovative, digitalized approaches. Here, we delve into the philosophy behind the PILOTING platform and the IVP, exploring both its raison d'être and its multifaceted offerings.

Historically, inspection processes were heavily manual, with many procedures executed on paper. This traditional method, while functional, is nonetheless ineffective, subject to human error, and poses certain challenges for maintaining consistency. The evolution to a digital approach not only mitigates these challenges but actively enhances the entire workflow. By guiding the inspection process from start to finish, the IVP empowers the PILOTING platform (see Figure 6.2) to ensure that crucial steps are not overlooked, offering a perfect blend of precision and productivity.

Harmonizing Diverse Data Stream

A significant challenge for asset owners is the reliance on third-party inspection service providers. Given that each provider may utilize different tools, robots, and data structures, the resulting data can be varied and non-homogeneous. Fortunately, the PILOTING Harmonization Layer adeptly tackles this challenge. Thanks to it, the IVP can offer the possibility to plan an inspection and visualize the collected

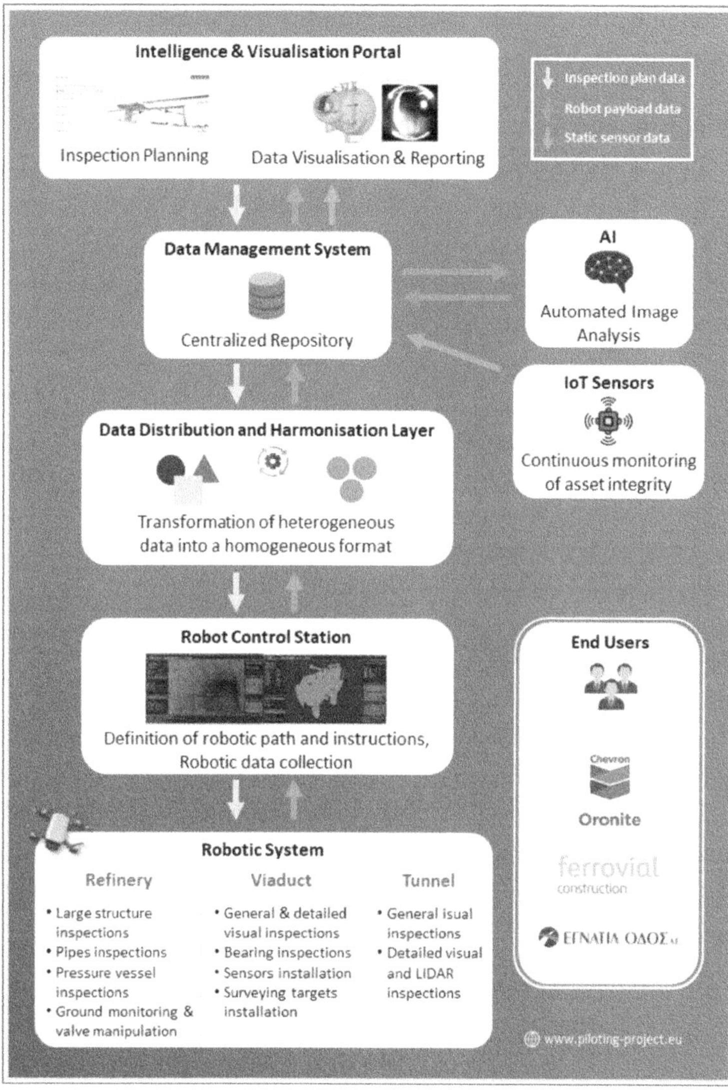

Figure 6.2. Overview of the IVP role within the PILOTING ecosystem.

data irrespective of the inspection service-provider proprietary systems. This ensures that asset owners can visualize a unified, comprehensible data set.

Holistic Planning

While the term 'planning' in inspection contexts often evokes mission strategies, it encompasses so much more. From securing the requisite permits and organizing logistics to deciding on personnel and equipment, planning is multifaceted.

The IVP simplifies this intricate process, ensuring that all the necessary files – from permits to personnel details – are organized and accessible.

Ownership, Compliance, and Choice

Asset owners, often lacking their inspection departments, are reliant on external service providers. The IVP platform can also empower them with the ability to solicit quotes from multiple inspection providers, review their plans, and select the one that aligns best with their requirements. Moreover, by housing data on-premises, asset owners can ensure both asset health and compliance with stringent regulations.

Data-Driven Paradigm

In the modern era, decisions anchored in data have a higher likelihood of success. Our platform facilitates data-driven decision-making, enabling forecasts of asset failures and facilitating proactive measures. This not only ensures asset longevity but also optimizes operational performance.

The Rise of the Digital Twin

The concept of digital twins – digital 3D replicas of physical assets – is gaining momentum, with corporations like Azure investing significantly in this domain. These digital assets offer unparalleled insights into their real-world counterparts, revolutionizing the way inspections and asset management are approached.

In sum, the PILOTING platform represents a paradigm shift in the world of inspections. It amalgamates the best of technology, innovation, and user-centric design to offer a solution that is profoundly impactful in optimizing asset health and longevity.

6.3 Design and Implementation

The essence of any effective system does not just lie in the objectives it intends to achieve, but crucially in the mechanics of how it aims to achieve them. In this section, we break down the intricate operational dynamics of our platform, offering a comprehensive view of its core functionalities and the philosophy that drives them.

Guided Inspection Workflow

At the heart of our platform is a meticulously structured inspection workflow. This is not a mere sequence of tasks but a harmonious integration of steps, each building upon the previous, ensuring that the inspection process is not just thorough but also practical. From the initial stages of asset identification to the culmination in data analysis and report generation, every phase is orchestrated to provide clarity, reduce redundancies, and facilitate accurate, actionable insights.

Asset-Centric Approach

The platform is fundamentally asset-centric. This means that every feature, tool, and function is designed with a singular focus: the asset being inspected. Such an approach ensures that all collected data, analyses, and subsequent actions are always contextually anchored to the asset in question. It fosters a holistic understanding, allowing inspectors, asset owners, and other stakeholders to have a comprehensive view of the asset health, needs, and potential areas of concern.

Advanced 3D Environment

Central to the IVP operational strategy is its emphasis on a 3D environment. While 2D data and traditional formats have their value, the shift to a 3D-centric approach brings several advantages. It facilitates a more intuitive understanding of spatial relationships, enhances visualization capabilities, and allows for a more immersive interaction with the asset data. By integrating this 3D focus, the IVP ensures that users can visualize and understand asset conditions with unparalleled clarity, bridging the gap between digital representations and real-world scenarios.

Portal Architecture

The IVP interface is designed as a modern web application, for easy access and delivery to end-users. The architecture of the portal is presented in Figure 6.3. At its core, the platform leverages specialized backend services dedicated to refining stored data (see Figure 6.3 – point-cloud utils app service and Image utils app service boxes). Think of these services as backstage technicians, expertly fine-tuning raw data saved in the Data Management System, whether it is converting complex point clouds or generating AI-annotated images. These services interact with the *IVP app service* through secure online communication channels (represented by arrows in Figure 6.3). Moreover, an added layer of a shared cache ensures that interim transformations, like specific data conversions, are stored systematically, speeding up subsequent processes (see Figure 6.3 – persistent caching box).

But the improvements are not just about rapidity. The platform underlying structure has also been fortified with enhanced security features. By streamlining its core operations, the platform has integrated advanced security measures, like VPN connections, directly into its main service. This not only ensures a safer environment for the data but also simplifies the overall system architecture, making it more resilient and reliable for users.

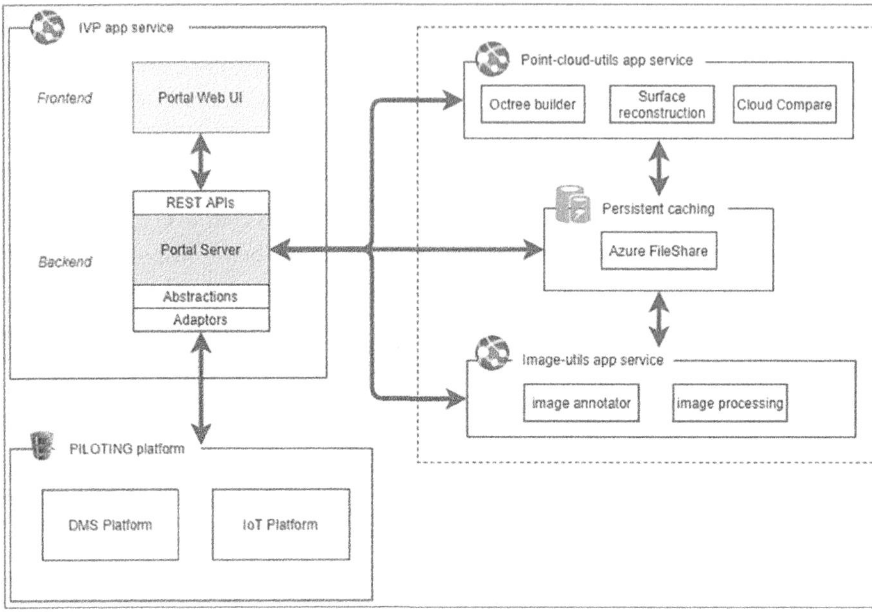

Figure 6.3. PILOTING application architecture. The IVP app service is a web application composed of a client and a server. The latter communicates with the PILOTING platform, responsible for data storage, and the point-cloud and image app services, responsible for data post-processing.

6.4 Workflow

In the evolving landscape of asset management, the introduction of digital inspection systems has marked a transformative shift toward enhancing precision and reliability. The integration of state-of-the-art technologies, ranging from 3D modelling to AI to advanced robotics, has ushered in an era of comprehensive asset analysis. Presented below is a structured outline of the sophisticated workflow (see Figure 6.4), meticulously crafted to facilitate precise inspection and comprehensive reporting:

1. **Create the Asset:** Understanding an asset intricacy begins with its digital replication, a Digital Twin – a precise digital 3D replica of the asset. By generating a Digital Twin, we establish a foundational reference point, allowing inspectors to interact with the asset virtually, even before the actual inspection commences. At this step, essential asset details are logged into the system, ranging from generic information to specific attributes tailored to the asset type. The robustness of the system permits the user either to incorporate an existing point cloud model or to sculpt one from scratch using the dedicated *AssetBuilder* tool (see 6.4.1.2).

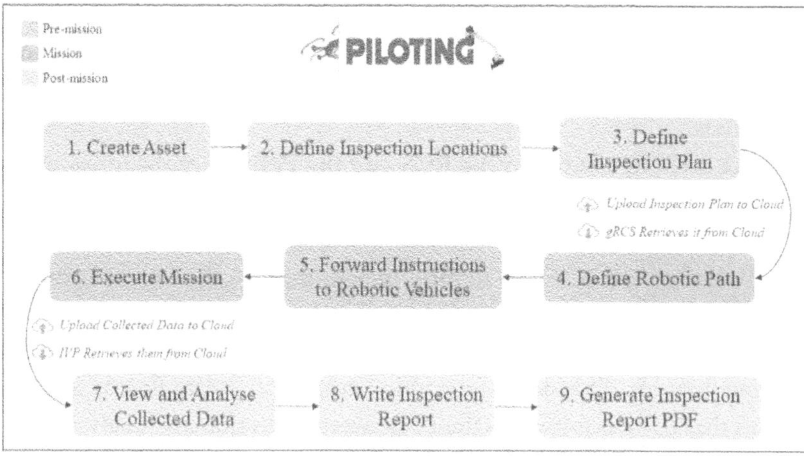

Figure 6.4. PILOTING inspection workflow.

2. **Identify the Inspection Locations:** Assets often possess areas more prone to wear and degradation. By pinpointing these critical zones, we ensure that the inspection process is proactive rather than reactive, anticipating problems before they escalate. Leveraging the 3D model, inspectors can judiciously select and define these areas, ensuring a focused and thorough examination.
3. **Define the Inspection Plan:** A well-laid plan prevents oversight and enhances correctness. Crafting a detailed inspection plan ensures a systematic approach, reducing potential gaps in the inspection and ensuring comprehensive coverage. First, a detailed roadmap is charted out, encompassing vital facets like the inspection duration and the scope. Second, the inspector uses the Inspection Locations defined on the 3D model in the previous step to plan specific tasks to perform at each location. The selection of the robotic system is also determined, ensuring the right tool for the right job.
4. **Data exchange between the IVP and the RCS:** Communication between the Inspection and Visualization Portal is essential for a successful operation. By facilitating this data exchange, we ensure that the robotic systems are equipped with the latest inspection strategies, operating optimally. Consequently, upon finalizing the inspection blueprint, the data is uploaded by the IVP to the cloud, ensuring accessibility and integration with the Robotic Control Station (RCS) platform. The latter retrieves the inspection plan, empowering the robotics operator to craft the optimal path for the robots, ensuring comprehensive coverage and precision.
5. **Forward Instruction to the Robotic Vehicles:** Precision in robotic operations hinges on clear instructions. By transmitting detailed directives to the robotic system, we set the stage for the inspection.

6. **Execute the Mission:** An asset true-health is revealed during this hands-on exploration. The robots, equipped with advanced sensors, delve into the asset, capturing a wealth of data that forms the backbone of the subsequent analysis.
7. **Data exchange between the RCS and the IVP:** Once the mission is completed, the collected raw data must be transferred from the robots to the IVP to make them available for analysis. These data are first transferred from the robots to the RCS, which in turn uploads them to the cloud. At this point, the IVP retrieves the mission data directly from the cloud and processes them to make them accessible and interpretable through the portal user-friendly tools. This direct interaction ensures that data flow is streamlined and that there is minimal latency between collection and analysis.
8. **Visualize and Analyze Data:** One of the standout features of the IVP is its suite of intuitive tools designed for data visualization. Inspectors are provided with a dynamic 3D environment, which not only showcases the data but also contextualizes it. This means that data points, images, or readings are not just presented as standalone entities. Instead, they are overlaid on the digital twin, offering spatial context. This immersive experience ensures that inspectors can correlate findings with their exact locations on the asset, making the analysis more insightful. Other important tools support the inspector during the data analysis process, such as the image annotator, the file explorer, the AI annotations validation, the IoT dashboard, and others. We will dive deeper into these tools in Section 6.4.5.
9. **Write the Inspection Report:** The culmination of the inspection process is the generation of a detailed report. Documentation is the key to actionable insights. Crafting a comprehensive report ensures that all findings, observations, and recommendations are systematically recorded, serving as a guide for future actions and decisions. The integrated Report Editor tool simplifies this task, providing inspectors with predefined templates and structures. Indeed, this tool is designed to offer a structured and intuitive environment for consistent documentation, ensuring that all critical aspects of the inspection are captured.
10. **Generate the PDF Inspection Report:** In an environment where multiple service providers operate, variations in report formats can cause confusion and inconsistencies. Generating a standardized PDF report, irrespective of the service provider, ensures uniformity in documentation. This standardization streamlines communication, interpretation, and decision-making across different stakeholders and organizations. Once the report is finalized in the editor, it is converted into a distributable format. The platform facilitates the generation of a comprehensive PDF, encapsulating

all the data, observations, and conclusions. Moreover, customizable templates maintain consistency and uphold the branding guidelines, ensuring the report integrity aligns with organizational standards. This PDF serves as a tangible record of the inspection, ready for sharing, archiving, or further analysis.

The fulfilment of these steps is a robust, detailed, and accurate reflection of the asset health, ensuring informed decision-making and timely interventions. This digital evolution not only enhances the quality of inspections but also propels the realm of asset management into a new era of precision and reliability.

In the subsequent sections, we will delve into a detailed exploration of each step, elucidating the intricacies and methodologies employed. This in-depth examination is aimed at providing a comprehensive understanding, ensuring you are well-equipped to harness the full potential of the system.

6.4.1 Create the Asset Digital Twin

In today's rapidly evolving industrial landscape, the concept of digital twins has emerged as a linchpin for effective operations and robust asset management. At its core, a digital twin is a dynamic digital representation of a physical entity or system (see Figure 6.5 for 3D representation). It provides a real-time look into the asset, capturing its current state, history, and even potential future scenarios.

The primary allure of digital twins lies in their ability to bridge the physical and digital realms. By replicating real-world assets in a digital environment, organizations can run simulations, predict failures, and optimize operational strategies without direct interventions on the actual asset.

For inspection processes, the value of digital twins becomes even more pronounced. They not only facilitate a comprehensive understanding of the asset current condition but also serve as an invaluable tool for planning and monitoring inspection missions. This proactive approach, enabled by digital twins, paves the way for timely interventions, minimizing risks, and ensuring the longevity of assets.

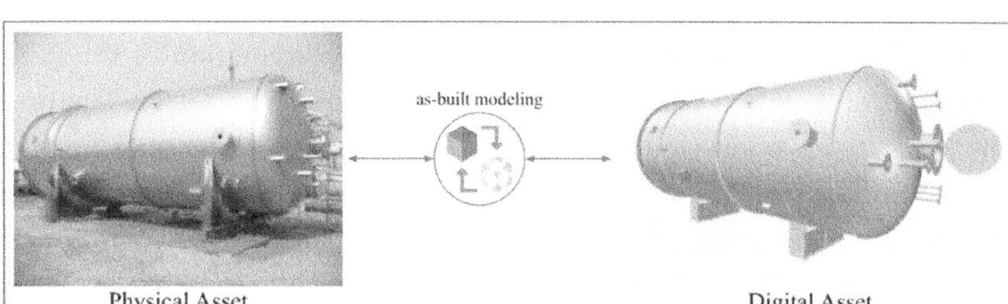

Figure 6.5. Example of a 3D digital twin. The digital asset was created with the Asset-Builder tool (see Section 6.4.1.2).

Furthermore, the integration of digital twins with modern inspection tools and methodologies allows for enhanced data visualization, leading to more informed decision-making processes. When an inspector views inspection data within the context of a 3D digital twin, the data becomes more than just numbers or images; it transforms into a story that paints a holistic picture of the asset health.

Consequently, a digital twin forms the bedrock of the inspection process:

- First, the inspector uses it to plan the mission, defining the areas to inspect and the tasks to perform on the real asset (see Section 6.4.3).
- Then, the robots use the 3D model as a reference to locate the real asset areas of interest and define a path to follow during the inspection.
- Finally, the inspector visualizes the data collected by the robots in a 3D environment, enhancing data visualization and analysis by making the information contextual and easier to understand.

Having underscored the significance of digital twins, it is essential to realize how they materialize within the Inspection and Visualization Portal (IVP). While the concept is vast and its implications profound, its actual implementation in the IVP is designed to be straightforward and user-friendly. The emphasis is the ease of use; indeed, the inspector can create a digital representation of the asset in a two-step process:

1. **Add asset information:** Enter relevant information about the asset itself. The data is categorized into generic information such as the asset name, unit, etc., and specific information based on the asset type. For example, a pressure vessel specific-information would include its orientation, internal diameter, and end type.
2. **Create/Upload the 3D digital twin:** Add a 3D model of the asset to create the digital twin. To do so, the portal supports both uploading a point cloud model or creating the 3D model from scratch using the *AssetBuilder* tool.

6.4.1.1 Point clouds uploader & viewer

A point cloud is a collection of data points. Each data point represents a specific point in space, collectively forming a detailed 3D representation of the subject, capturing its shape, volume, and even color or texture in some cases (see Figure 6.6).

The necessity for point clouds arises from the need for precise spatial information. In industries like construction, infrastructure maintenance, or architectural conservation, understanding the intricate details of a structure or environment is paramount. Point clouds provide this detailed insight, offering a comprehensive digital representation that can be analyzed, manipulated, and used for various purposes, from structural assessments to virtual simulations.

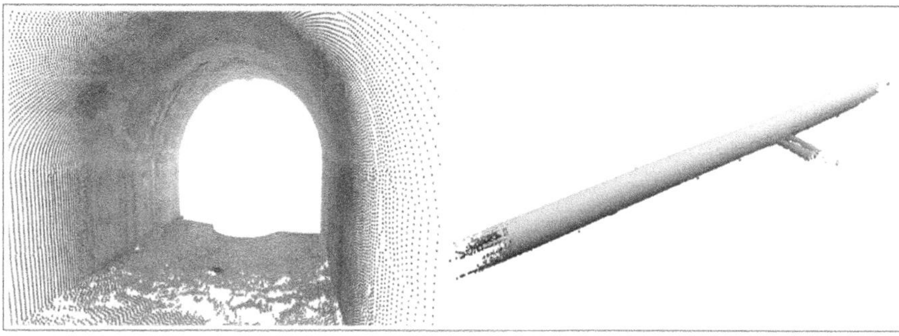

Figure 6.6. Inside (left) and outside (right) view of a tunnel point cloud. Captured with a LIDAR scanner by one of the PILOTING robotic systems and visualized with the Point Cloud Viewer.

Generating a point cloud typically involves the use of advanced scanning equipment. Devices such as LIDAR (Light Detection and Ranging) sensors or 3D laser scanners capture thousands, sometimes millions, of data points by emitting laser beams and measuring the time it takes for the beam to return after hitting an object. The data collected is then processed and translated into the point cloud.

Consider, for instance, the inspection of a viaduct or tunnel. Given the vastness and complexity of such structures, a manual inspection can be time-consuming and potentially miss critical details. By employing a robotic vehicle equipped with LIDAR sensors, one can easily scan the structure. The resulting point cloud provides a detailed 3D representation of the viaduct or tunnel, highlighting potential areas of concern such as cracks, deformations, or wear. This digital twin of the real-world structure then serves as a foundation for further analysis, maintenance planning, or even simulation of potential interventions.

Following the generation of this detailed 3D representation, the next logical step is to make this data accessible and usable. In many use cases, a 3D map of the surroundings is generated during a preliminary mission, either by a robotic total station or 3D LIDAR sensors mounted on the robotic vehicles. The resulting point cloud can be uploaded to the IVP using the Point Cloud Uploader tool, which is designed to be compatible with industry-standard formats, ensuring interoperability with other software solutions.

Upon uploading the point cloud to the portal, the Point Cloud Viewer enables users to effortlessly visualize vast and intricate point cloud datasets from remote storage. This is achieved through an adaptive loading algorithm, ensuring a smooth and efficient user experience regardless of the data size.

This environmental representation is crucial for localization, intelligent perception, and autonomous navigation of both aerial and ground robots. The digital twin thus morphs into an essential part of the inspection process, presenting a comprehensive, interactive, and highly informative depiction of the physical asset.

6.4.1.2 AssetBuilder tool

In the field of robotic system inspections, the choice of digital representation can vary based on the technical specifications of the robotic system and the use case at hand. While capturing point clouds is a popular method, it may not always be the most suitable approach. For this reason, the IVP also supports a compact alternative to create digital twins: mesh models. Mesh models come with the advantages of quicker creation times and reduced file sizes, making them particularly beneficial when the robotic system necessitates uploading the digital twin to its navigation framework.

To cater to this need, the Portal integrates the AssetBuilder tool (see Figure 6.7: Example of a piping rack created with AssetBuilder.), an interactive drag-and-drop modelling tool designed to quickly create mesh models of assets from basic parts. The tool offers similar features to other CAD programs but with a focus on specific asset classes. It provides pre-designed highly configurable parts based on industry standards.

During the model assembly, compatible parts automatically stick together, preventing incompatible configurations, and making it user-friendly even for non-experts. The AssetBuilder tool supports pressure vessels, piping, and storage tank assets.

Delving into its functionalities, users can:

- Interactively engage with the 3D model by selecting, dragging parts, or altering the viewing angle.
- Assemble assets piece by piece. For instance, when modeling a pressure vessel, one could start with the shell, then add a head, nozzles, and so forth.

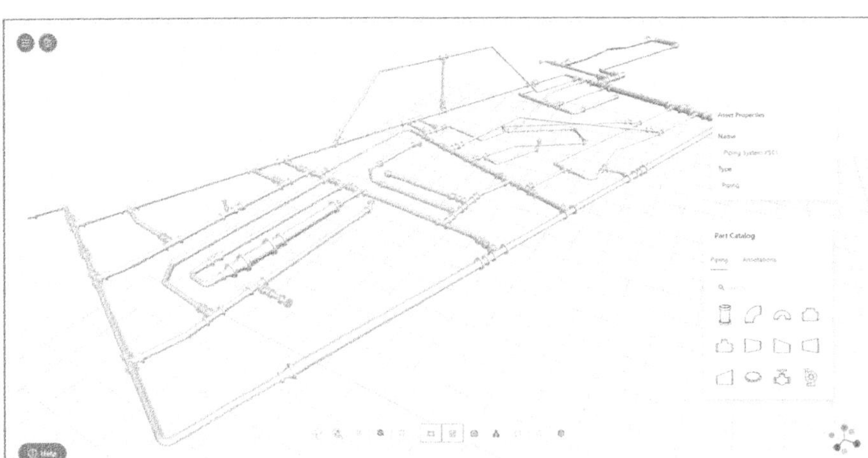

Figure 6.7. Example of a piping rack created with AssetBuilder.

Similarly, a piping model could begin with a pipe cylinder and progress to include elements like elbows and pumps.
- Customize each part extensively by adjusting its parameters, ensuring that the 3D representation mirrors the real asset specifications and dimensions accurately.
- Capture and save specific views or details using the screenshot feature.

All this considered, the AssetBuilder tool amplifies the versatility of the Portal, allowing for more diverse and accurate digital twin representations, vital for precise robotic inspections.

6.4.2 Define Inspection Locations

Following the creation of the digital twin, the next crucial step in the digital inspection process is determining the inspection locations. These are certain regions of an asset that, due to their susceptibility to damage like corrosion or bearing wear, require regular inspection. Some of these locations may be known at the asset design stage, while others can be identified during the asset lifecycle, often as a result of previous inspection findings.

Inspection locations, as the name suggests, are not monolithic entities. Depending on the specific use case and the nature of tasks or measurements to be performed, they can vary significantly in terms of their scope and specificity.

For structures like viaducts, the focus might be on expansive regions such as the deck or the pillars. Given their vastness, these are typically called "**inspection areas**". They represent larger segments of an asset that demand thorough examination due to their critical role in the structure integrity (see Figure 6.8 – right).

On the other hand, when considering assets like pressure vessels, the attention often narrows down to specific components like nozzles, and so on. Such distinct and defined components are termed "**inspection parts**". These parts, though

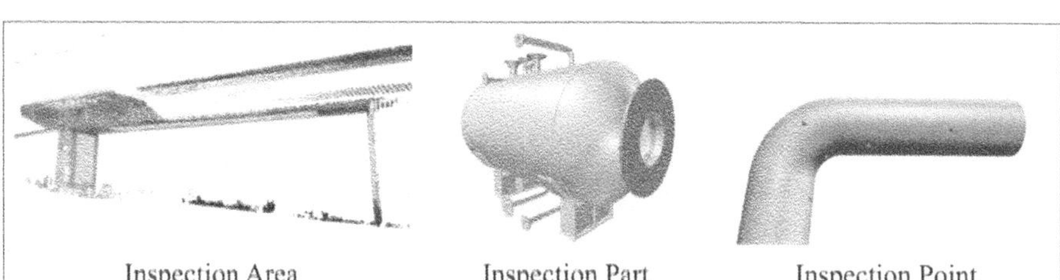

Figure 6.8. Examples of different inspection locations on the IVP: pillars and deck of a viaduct, depicted as bounding boxes on the asset surface (left); the nozzle of a pressure vessel, highlighted in orange (center); specific points of a pipe, depicted as small spheres (right).

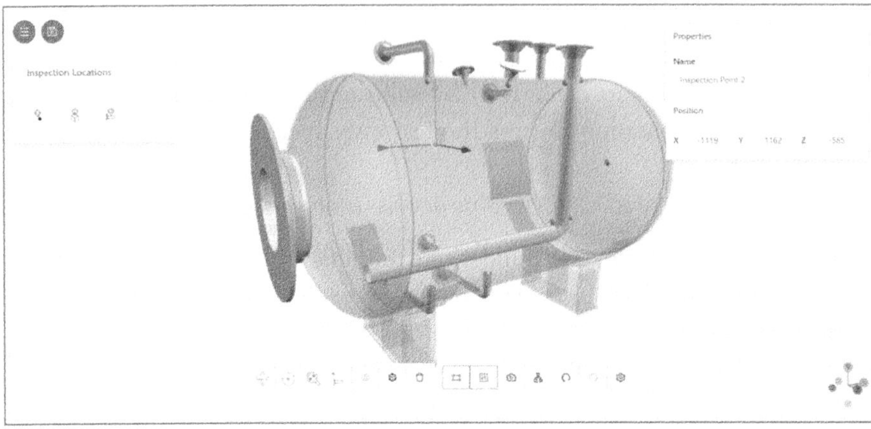

Figure 6.9. Interface to manage inspection locations directly on a digital twin. The inspector can add new inspection locations from the "Inspection Locations" panel, as well as edit existing ones by interacting with the 3D model.

smaller in scale compared to inspection areas, are equally pivotal and have their own unique inspection criteria (see Figure 6.8 – center).

Moreover, there are instances where the emphasis shifts to very specific points. For example, during Ultrasonic Thickness (UT) inspections, the interest is shifted to understanding the asset thickness at very precise points. These are not broad areas or distinct parts but rather precise locations, aptly named "**inspection points**" (see Figure 6.8 – left).

Having established the conceptual framework of inspection locations, it is pertinent to note the transformative role of the digital twin in this paradigm. The 3D model not only serves as a digital representation but also as a canvas upon which the inspection locations can be defined. The digital twin, in essence, provides a one-to-one mapping, ensuring that each location defined on it corresponds accurately to a real location on the actual asset.

The IVP provides an intuitive interface where the inspector can interact with the digital twin to create, modify, and visualize the inspection locations (see Figure 6.9).

Once the inspection locations are created, the portal adds them to the asset inspection location list and shows them on the 3D model. The inspector can always add new locations to adjust the inspection strategy based on new findings or changing conditions. The inspection locations can be reused across multiple inspection plans, promoting consistency in the inspection process.

6.4.3 Define an Inspection Plan

With inspection locations firmly established, the emphasis shifts to the formulation of a comprehensive inspection plan. Such a plan is vital as it lays down the strategic blueprint for the entire inspection mission. While the inspection locations identify 'where' to inspect, the inspection plan elucidates *how* the inspection will be conducted. This includes details about the types of inspections, the sequence, and the specific robots and tools to be deployed. Ensuring that this plan is thorough and well-defined is critical, as it becomes the primary reference for robotic operators and inspectors alike, guiding them through the field mission.

Transitioning from the broad overview of the process, the portal user interface guides inspectors from one step to the next. The platform, designed with clarity and efficiency in mind, ensures that the conceptual steps are mirrored in its operational design. As inspectors embark on the task of defining the inspection plan, they are greeted with intuitive tools and prompts on the platform, which facilitates the input of essential details. This user-centric design ensures that while the inspector focuses on the critical aspects of the plan, the platform handles the intricate details, ensuring data integrity and ease of use.

6.4.3.1 Adding inspection plan information

Formulating the inspection plan begins with the inspector inputting general details such as the plan name, its start and end dates, and a concise description outlining its scope. This sets a clear timeline for the inspection process and outlines the overall objective of the plan.

Furthermore, to ensure a comprehensive and safety-compliant inspection, various files such as risk assessments, safety regulations, and emergency procedures can be attached to the plan at this stage. These attachments serve as important reference materials during the inspection mission and provide additional context for the tasks to be performed.

6.4.3.2 Establishing inspection tasks

Before delving into the intricacies of the platform, it is essential to understand the fundamental concept of an "**Inspection Task**". In the field of asset inspection, a task can be visualized as a specific action or a series of actions that a robot or inspector needs to perform on an asset such as an Ultrasonic Thickness (UT) measurement, a cleaning procedure, or a visual inspection. For instance, imagine a large viaduct, over time, natural elements like rain, wind, and temperature fluctuations might have affected its structural integrity. To assess this, an inspection task might involve a drone flying over to capture high-resolution images of the viaduct pillars, looking for cracks or erosion. Another task might involve a ground robot equipped with sensors, moving along the bridge length to measure vibrations, ensuring the bridge stability. Each of these actions, from visual image capture to vibration measurement, represents an individual task in the broader spectrum of the inspection process.

Building on this understanding, consider the scenario of a robotic vehicle inspecting a large-scale refinery unit. This robot is not just capturing visual data. First, it might deploy a brush or air jet to clean a specific section of the infrastructure. Only after ensuring the area is free from debris or dust does it proceed to visually inspect the surface for any signs of wear or damage. Following this, the robot may then engage its UT sensors to evaluate the thickness of the material, checking for any internal damage or erosion. This sequence is imperative. If the robot were to conduct a UT inspection on a dirty surface, the results might be compromised. In real-world scenarios like this, the order in which tasks are performed is crucial.

Just as tasks in real-world scenarios need to be executed in a specific sequence, the portal design acknowledges and encapsulates this reality. It allows the user to define tasks in a sequential manner that mirrors actual operational needs, ensuring that automated path planning is optimized. Once the plan details are set, the 3D editor interactive-mode allows the inspector to engage with the digital model to assign specific inspection tasks to the previously defined inspection locations.

By selecting an inspection location, the inspector can assign the task that the robot is to execute at that location (see Figure 6.10). Notably, at a designated inspection location, multiple tasks can be orchestrated in a particular order. Whether it is cleaning first, then a visual scan, followed by a UT assessment, each step can be clearly delineated. Users can add, remove, or adjust the sequence of these tasks. Moreover, to streamline this process and enhance user experience, every inspection type is color-coded, facilitating quick and easy task identification.

Figure 6.10. Example of the inspection tasks 3D editor. Inspection locations allow task additions with few simple clicks, turning them green once tasks are assigned. The blue arrows indicate the robot inspection-sequence: starting with the left pillar, moving to the deck, and concluding with the right pillar.

6.4.3.3 Selecting the robotic system

With the plan information and inspection tasks defined, the final step involves selecting the robotic system that will execute the inspection tasks.

Each robot is tailored to cater to specific inspection tasks and use case scenarios. From confined spaces to expansive industrial structures, each setting poses unique challenges that demand specialized robotic capabilities.

Consider the inspection of pressure vessels, which are intricate structures that operate under high pressure and can pose significant risks if not properly maintained. For such assets, a generic robot might not suffice. Instead, a specialized one, equipped with the ability to magnetically adhere to the vessel surface, is vital. This not only ensures a stable inspection process but also enables the robot to traverse vertical or even inverted surfaces.

Therefore, the inspector needs to choose a robotic system that aligns with the specific needs of the inspection. Once the robot has been selected, the process progresses as elaborated in the next section: the inspection plan with its detailed tasks is archived in the cloud and the data is transferred to the Robot Control Station, setting the stage for the actual inspection process.

6.4.4 Data Exchange Between IVP and RCS

In the complexity of asset analysis, data streams play a key role in ensuring accuracy and timely decision-making. The ability to move, process and interpret data quickly and comprehensively is critical. In the given scenario, the data exchange is designed as a bridge between the platform and the robotic systems, ensuring the integration of processes and technologies.

Once the analytics process is carefully designed and completed, the platform quickly uploads data to the cloud. This cloud-based storage strategy assures that data will remain accessible and primed for integration with the Robotic Control Station (RCS). Once uploaded, the RCS retrieves the inspection plan (see Figure 6.11), which then becomes the guideline for the robotics operator. With this data in hand, the operator delineates the most encompassing path for the robots, ensuring they cover every essential location with precision.

After the mission, the robots, having gathered a vast amount of data from images to specialized measurements, transfer this raw information to the RCS. Then, the later uploads these data sets to the cloud (see Figure 6.11), This is where the IVP steps in, retrieving the mission-specific data directly from the cloud repository. The portal processes this raw data, molding it into formats that are both accessible and interpretable through its suite of visualization tools. This direct interaction between the RCS and the IVP ensures not just a streamlined flow of data but also guarantees that there is a minimum latency between data collection and its subsequent analysis.

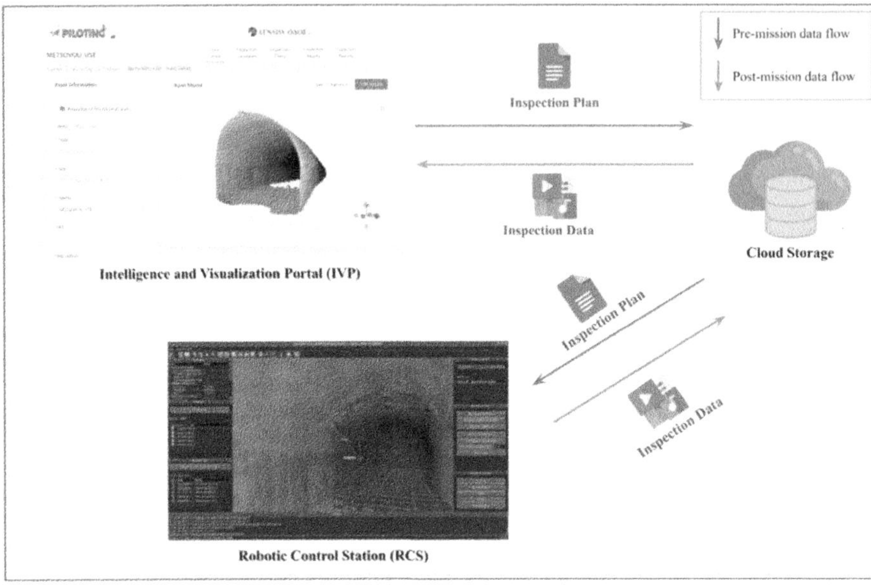

Figure 6.11. IVP and RCS communication, pre and post mission data exchange.

The beauty of this system lies not just in its technological excellence, but also in its ability to mirror real-world processes and needs into a digital realm, making the entire inspection cycle accurate and timely.

6.4.5 Data Analytics Tools

During the mission execution, the robotic systems gather crucial data such as images, videos, and UT measurements. Once the inspection is completed, the collected data is then forwarded to the portal as explained in Section 6.4.4, processed, and made ready for analysis.

The data analysis step is pivotal in the inspection process. While raw data, in isolation, holds value, its true potential is unlocked when it is systematically explored and contextualized. Analyzing data discerningly can unearth insights that are crucial for maintenance decisions, risk assessment, and future inspection planning. This section explores the assortment of tools the portal provides for data analysis and enhanced visualization.

The portal's suite of data analytics tools offers a convenient means to explore and visualize inspection data within the context of the 3D asset model. Inspectors can effortlessly access all collected data, examine them, and export or save them for integration into a report or further scrutiny. The Inspection Result Page on the portal serves as the hub where all the collected data can be navigated and viewed directly.

6.4.5.1 Inspection data in 3D context

Understanding the data in its raw form can be challenging. For instance, imagine you are looking at an image captured by a drone during a viaduct inspection, it showcases signs of early moisture. Is the picture taken in a critical area where further investigation is required or on a surface that is inconsequential? Determining this becomes difficult without the necessary context.

A key goal of the portal is to present inspection findings within the context of the inspected asset, moving beyond the raw data by focusing on a user-friendly and easily comprehensible visualization. Indeed, the IVP elevates data analysis by offering the ability to analyze collected data within a 3D context for various assets, such as refineries, viaducts, and tunnels. This immersive feature equips inspectors with a comprehensive understanding of the data, facilitating a more intuitive and engaging interpretation.

Delving further into the data analytics tools, we explore how the portal enhances the visualization of robot trajectories, inspection images, and Ultrasonic Thickness (UT) scan measurements, providing the inspector with a wealth of actionable information.

Trajectory

Consider a tunnel inspection performed by a ground robot. Without trajectory data displayed in relation to the tunnel 3D model, the user is left with a myriad of questions: – did the robot traverse the entire length of the tunnel? – was there a specific section where it lingered longer, possibly indicating a problematic area? – was the robot path in line with the planned route?

Similarly, envision an aerial inspection of a viaduct using a drone. The trajectory data, when juxtaposed with the 3D representation of the viaduct, can offer valuable insights. If the drone trajectory indicates prolonged hovering over a particular section, it suggests that this area might require special attention, possibly due to visible defects or anomalies. On the flip side, the trajectory can also serve as a validation tool. It confirms that the drone comprehensively covered the intended regions, providing confidence in the inspection thoroughness. Moreover, any deviations from the planned path can be identified – like if the drone inadvertently entered restricted airspace.

Such contextual visualization not only enhances the comprehensibility of inspection activities but also fortifies the decision-making process, allowing inspectors to act promptly and accurately based on the robot movements and actions during its mission.

To achieve this, all mission results encompass a localization-telemetry file generated by the RCS in a uniform format for all the robotic vehicles. This file carries

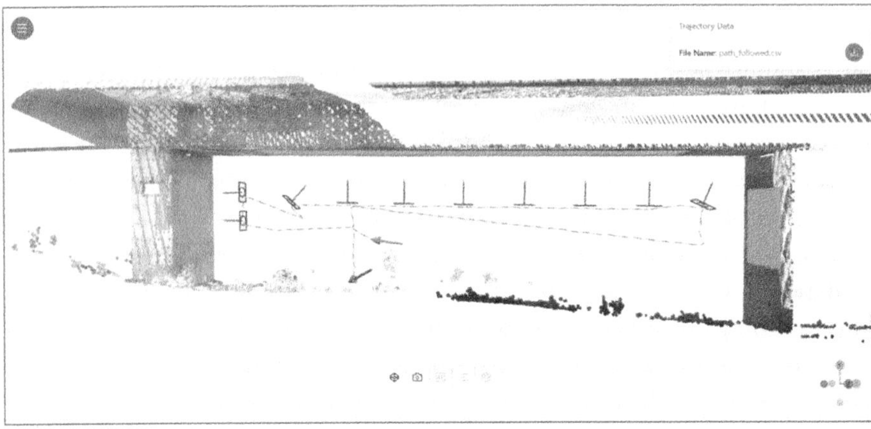

Figure 6.12. Drone trajectory overlaid on a digital twin of a viaduct, depicting the exact flight path the drone undertook during the inspection.

information about the position, orientation, speed, and other parameters of the robotic vehicle relative to a local 3D reference map during the mission. This data provides a detailed record of the robot path during the inspection, which can provide valuable context for understanding the collected inspection data.

The portal processes this data and visualizes the inspection trajectory in the context of the asset 3D model. In this way, the trajectory file is mapped onto the 3D representation of the asset. This allows the inspector to answer all the questions previously discussed and see exactly how the robot movements relate to the structure of the asset itself.

Such visualization is crucial to verify whether all vital areas of the asset, including those hard to reach, have been properly covered during the mission. This information is invaluable for the inspection report and provides a comprehensive understanding of the inspection mission extent.

Figure 6.12 shows an example of a drone trajectory overlayed on the digital twin of a viaduct.

Images

While a picture speaks a thousand words, in the realm of asset inspection, without context, those words can become almost useless. Consider a visual inspection in a tunnel, the sheer volume of images produced during the inspection can present a real challenge: – where exactly were these images taken? was it near a known vulnerability or is it a new one? Without a clear frame of reference, it can be hard to contextualize them with respect to the overall asset. Indeed, inspection images usually appear as a disconnected assortment of snapshots, making it hard to navigate through them and analyze the asset condition adequately.

Figure 6.13. 3D visualization of inspection images. Blue markers with arrows indicate the specific asset regions represented by each image, while the violet highlight identifies the image currently displayed in the "Properties" card.

This is where the portal 3D viewer comes into play. By leveraging the metadata associated with each image, the 3D viewer can position the inspection images within the context of the asset model. This situates the images within a clear, intuitive framework that mirrors the actual structure of the asset, transforming what was once a confusing array of isolated snapshots into a coherent, navigable 3D landscape (see Figure 6.13).

When these images are integrated with the asset 3D representation, inspectors can accurately determine the specific location of each image, weaving together a clear visual story. For instance, this approach allows for the identification of a corrosion mark on a digital representation of a pipeline and facilitates the swift determination of its exact location on the actual infrastructure.

On the digital twin, inspection images are represented by blue arrows, indicating the position and orientation of the camera where the image was taken. By simply clicking on them, the inspector can visualize the image taken at that precise point of the asset.

For certain use cases such as internal inspections of a pressure vessel, images are presented on the model with a distinct representation, taking into account the confined space. This tailored visualization approach enhances the exploration and understanding of the images even in smaller models than viaducts, tunnels, and tanks.

This innovative approach to image visualization provides a solution to the problem of data overload commonly faced in visual inspections. By contextualizing the images within a 3D framework that mirrors the asset actual structure, the portal greatly enhances the inspector's ability to analyze the collected data, facilitating more accurate and insightful interpretations.

UT scan measurements

When assessing the integrity of pressure vessels, pipes, or storage tank walls, knowing the wall thickness at various points is crucial. A decrease in thickness could be indicative of wear, corrosion, or other structural vulnerabilities. However, the raw data from these UT scans can be complex and difficult to interpret, especially when dealing with large areas or multiple points of measurement. Without a clear visual reference, it can be challenging to understand how the readings relate to the physical structure of the asset and to identify areas of concern such as potential thinning or corrosion.

Again, the portal overcomes these challenges with its 3D Viewer, which has the capability to display UT scan data directly onto the digital twin. Depending on the inspection scenario, UT readings can be taken at a single point, along a line, or over an area on the asset surface (see Figure 6.14 – top left).

The 3D Viewer automatically converts UT measurements into an interactive C-Scan thickness scan (or corrosion map) and overlays it onto the model at the

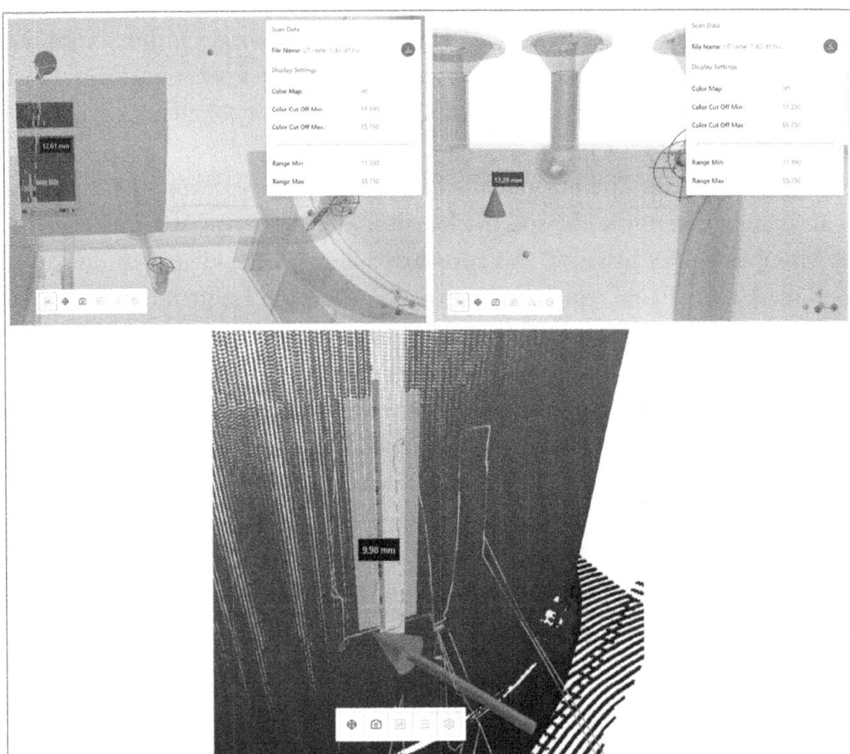

Figure 6.14. UT measurement visualizations for diverse inspection scenarios: area-based readings on the asset surface (top left); precise point-specific readings (top right); and linear measurements ideal for extensive structures (bottom).

location of the defined inspection area. This 3D visualization of UT data allows inspectors to contextualize each measurement on the asset, providing a clear visual reference that facilitates the interpretation of the data. Areas of potential concern, such as thinning or corrosion, can be easily identified and tracked across the asset.

The generated corrosion map can be customized to adjust the visualization as needed. For example, the adjustment of cut-off values (min/max) allows for tailored data analysis, while the different color map option (jet, rainbow, viridis, etc.) improves the visualization of the scan data (see Figure 6.14 – top right).

To enhance the usability and interpretability of the UT scan data visualization, the 3D Viewer provides the possibility to interact with the corrosion map. Positioning the cursor over a point on the corrosion map instantly displays the specific thickness measurement taken at that point. This instant access to data eliminates the need for inspectors to sift through a separate data table or document to locate corresponding measurements. Indeed, the feature significantly boosts the inspector ability to engage with the data, making understanding and analysis a more streamlined and intuitive process (see Figure 6.14 – bottom).

Overall, these tools offer the inspector a highly intuitive and user-friendly interface to visualize and analyze inspection data. They enable a more comprehensive understanding of the asset condition, facilitating decision-making, and potentially identifying issues before they escalate into more serious problems.

6.4.5.2 Image annotation & AI annotations

Nearly all use cases will yield a wide range of inspection images, which can be opened and visualized directly on the portal. By clicking on an image name or its preview, the image viewer opens in full-screen mode. This immersive view equips the inspector with an array of features such as zoom, pan, rotate, preview, and navigation to the next image.

Imagine a scenario where an inspector discovers a corrosion spot on a pipeline. Without annotating the image, other team members might overlook this small, yet critical detail. However, with a clear annotation marking the spot and detailing the observation, the risk of oversight is significantly reduced.

Having a dedicated tool for image annotation amplifies the benefits even further. It ensures consistency in communication, aids in quicker data interpretation, and streamlines the process of report generation. Additionally, the integration of Artificial Intelligence (AI) takes this a step further. AI-driven annotations can automatically identify and categorize common defects, significantly reducing the time inspectors spend on manual evaluations and ensuring that no critical details are missed. In essence, the fusion of manual and AI annotations promises a comprehensive and precise inspection process.

The subsequent sections delve further into these tools.

Image annotator tool

Is common that specific areas in an image scream for attention. Consider a scenario where a bridge deck reveals a hairline crack in a visual inspection. Without a clear marker, this critical detail could be overlooked by stakeholders reviewing the image.

The Annotation tool (see Figure 6.15) transforms an image from a mere visual representation to a canvas of insights, providing tools to clearly point out areas of concern or interest. Such annotations become especially vital when conveying findings to diverse audiences, ensuring clarity and mitigating ambiguity. Indeed, annotations can serve as a form of visual communication, making it easier to convey findings to others who may be reviewing the inspection results, such as team members, clients, or regulatory authorities.

The annotation tool offers a comprehensive suite of editing features:

- **Bounding Boxes:** Rectangles and ellipses can be added to focus attention on a specific area of the image. For example, during a bridge inspection, they can be used to highlight potential cracks or wear on the structure pillars.
- **Arrows:** Used to point out and highlight findings. For example, when inspecting a refinery unit, arrows can pinpoint a specific leakage area or a corroded section of a pipeline.
- **Point Marks:** Highlight specific points of interest or concern. For example, in the case of a large storage tank, point marks can emphasize small dents or imperfections that might be overlooked.
- **Text:** Additional notes or comments can be inserted directly into the image, providing immediate context, and reducing the need for extensive supplementary notes.

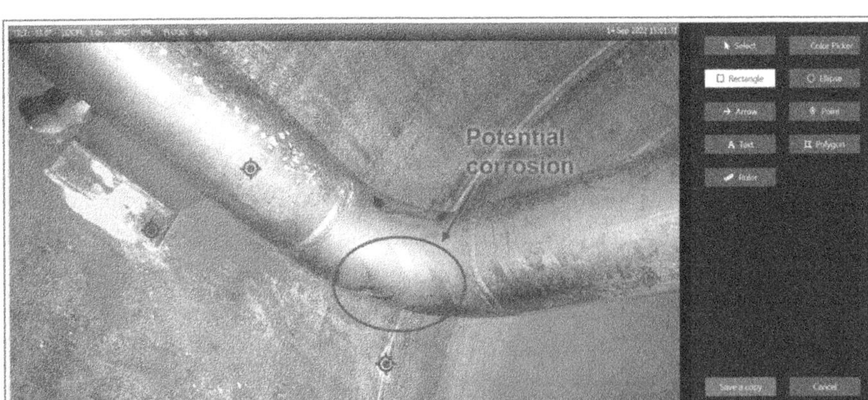

Figure 6.15. Image Annotator Tool showcasing different annotations: ellipsis, arrows, text, points, different colors.

- **Custom Shapes:** Draw custom shapes to outline irregular areas. For example, in a tunnel inspection, custom shapes can be drawn to trace the contours of irregular water seepage patterns or areas affected by mold.
- **Color Palette:** Allows users to choose from a variety of colors for their annotations.
- **Measurements:** This tool helps inspectors accurately measure the dimensions of any damage by simply drag-and-dropping the tool on the image.

The original images are never altered: a separate annotated image is generated, enhancing the inspection report with a detailed visual reference.

In conclusion, the Image Annotator Tool remains central to the inspection process, combining visual information with insightful annotation. As the inspection landscape evolves, tools like this will be indispensable in bridging the gap between raw data and actionable insights.

AI annotations

The huge volume of visual data produced during the inspections can be overwhelming. Manually sifting through each image to identify signs of deterioration would be not only time-consuming but also error-prone. Artificial Intelligence (AI) steps in as a game-changer, streamlining this process. Through AI-assisted annotations, potential defects are automatically identified, classified, and annotated, ensuring that inspectors are immediately directed to areas that require their attention, enhancing accuracy in the analysis.

The measurement tool of the annotation editor works in conjunction with the AI system to provide users with precise measurements of damage based on the AI-made annotations. Additionally, users can edit the AI-generated measurements and manually add new ones to the images, ensuring the annotations are as accurate as possible.

To facilitate the analysis, the IVP has introduced a color-coded system for the AI annotations. Each prediction is assigned a different color, making it easy for the inspector to distinguish between different predictions (see Figure 6.16).

However, AI predictions could be inaccurate, so it is crucial that the inspector validates the AI-generated annotations. To support this important step, the portal provides a list of all annotations along with their corresponding colors. Here, the inspector can mark the AI-generated annotations as correct or incorrect, ensuring accuracy, reliability, and fairness through human evaluation of the AI predictions (see Figure 6.16 – right panel).

In summary, the Image Annotator Tool, enhanced by the AI model, greatly enhances the ability of inspectors to interact with and analyze inspection images, leading to more accurate and effective inspections and communication.

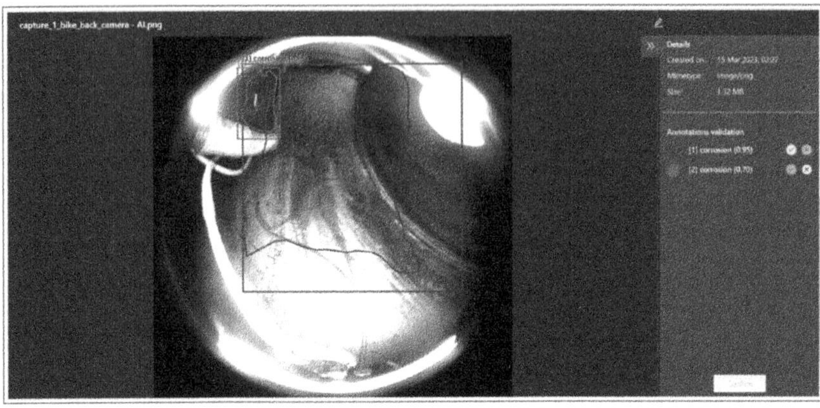

Figure 6.16. AI annotations tool: annotated image and annotations validation panel (on the right).

6.4.5.3 File management

Inspections produce not just vast amounts of data but also a large number of associated files. From technical drawings of the asset and inspection risk assessments to permissions and safety documentation, the volume and diversity of these files are significant. Indeed, efficient file management is not just a convenience, it is a necessity.

The primary rationale for robust file management is to ensure consistent operations. For instance, amid the myriad of files, it is vital to swiftly pinpoint specific documents, be it a safety protocol or a technical schematic of the asset. Moreover, with the plethora of mandatory procedures and regulations governing inspections, a well-organized system aids in verifying that no critical document is amiss. This includes essential permissions, safety documentation, and other regulatory requisites, ensuring that inspections are not just thorough but also compliant.

For asset owners, the significance of file management takes on an added dimension. The ability to readily access previous inspection reports facilitates the creation of a comprehensive asset history. This historical perspective can be invaluable in tracking the asset health over time, understanding its wear patterns, and pre-emptively identifying potential issues. Additionally, having immediate access to assets' manuals, blueprints, and security protocols ensures that any intervention, be it maintenance or a more intricate repair, is informed and accurate.

To this end, the IVP offers an intuitive interface designed to make information not only available but also easily findable, accessible, and organized (see Figure 6.17). The system simplifies the maintenance process by providing an environment in which each file has its own place and can be retrieved with minimal effort. This guarantees that professionals can focus on the core task at hand: ensuring the health and longevity of the asset.

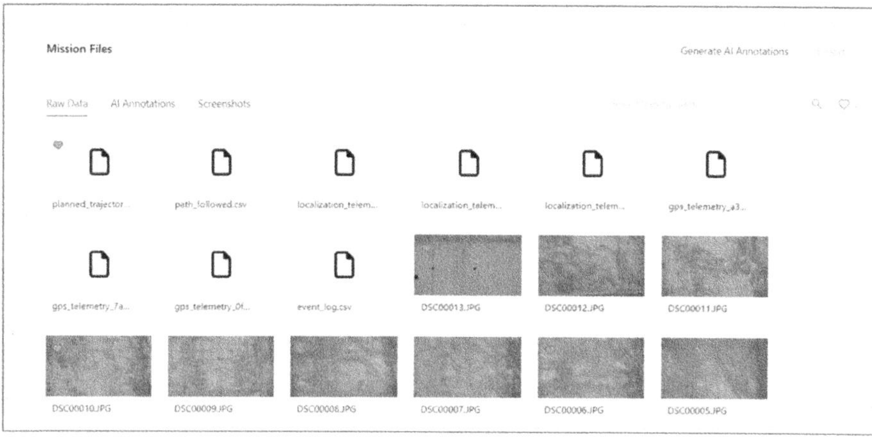

Figure 6.17. File Management for data gathered during the mission execution.

6.4.5.4 IoT dashboard

As technology advances, the integration of static IoT sensors into inspection procedures has provided an innovative method to continually monitor asset health.

For instance, consider a large bridge subjected to daily heavy traffic. By placing vibration sensors along critical junctions and support beams, it becomes possible to detect unusual movements or shifts that could be early indicators of structural problems. Similarly, in industrial setups like chemical plants, temperature sensors can alert operators about potential overheating issues, preventing accidents and costly downtimes. Therefore, these sensors serve as the asset continuous health monitoring system, providing real-time data that can preemptively indicate issues before they escalate into significant challenges.

In the initial pilots for the refinery and viaduct use cases, a multitude of these sensors were strategically placed at select asset locations. Their primary objective is to monitor vibrations, aiming to pinpoint anomalies that might necessitate a comprehensive physical inspection of the asset. To display in an intuitive way the data collected by these sensors, the IoT Dashboard has been intricately designed.

The IoT Dashboard (see Figure 6.18) is a hub of data, integrating with the SensorThing database to provide up-to-date insights from various sensors. This system prioritizes accuracy, ensuring asset owners and inspectors receive reliable information for decision-making directly on their laptops.

Users have access to a detailed list of sensors, which provides essential insights at a glance. Each sensor entry is accompanied by a visual representation, making identification straightforward. Furthermore, information about the sensor type (e.g., vibration, temperature) and associated data streams are readily available, giving users a holistic understanding of each sensor function and performance.

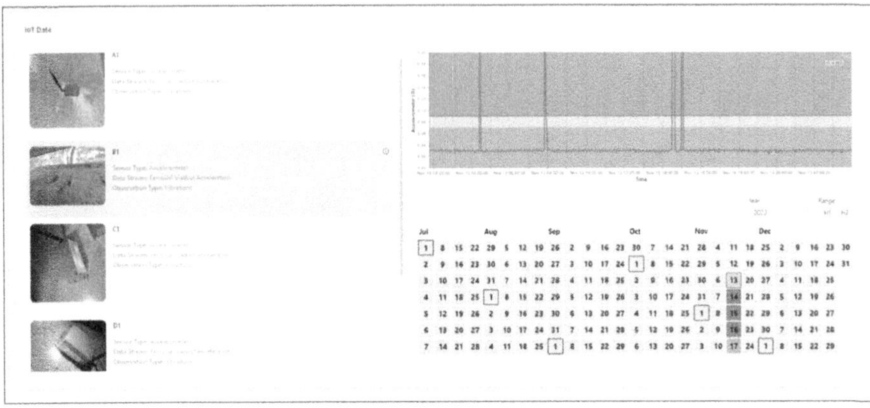

Figure 6.18. IoT Dashboard.

Upon selecting a sensor, users are treated to a vivid display of its data. Measurements are elegantly plotted on a line graph, with time as the reference point. This data is further enhanced with a color-coded system to instantly communicate the health status of the asset, helping users quickly discern potential problem areas:

- **Green (safe):** indicates that the measurements are within acceptable limits.
- **Yellow (warning):** suggests potential issues or readings nearing unacceptable levels.
- **Red (critical):** signals readings that are beyond safe limits and might require immediate attention.

Below the primary data visualization, users can delve into historical data through a calendar heatmap. Each day is represented as a color-coded square, consistent with the aforementioned coding system, allowing for easy navigation through historical data. By selecting specific dates, users can zoom into data from particular periods, enabling focused analysis and understanding of trends over time. This design empowers the scanning of the accumulated data, allowing inspectors to focus attention on specific dates or periods that demand a closer look.

In essence, the IoT Dashboard, with its visual aids, ensures that inspectors and asset owners can effortlessly monitor, analyze, and make informed decisions. It transforms the traditionally cumbersome process of data interpretation into a streamlined and user-friendly experience, underscoring the importance of technology in modern inspection procedures.

6.4.5.5 KPI dashboard

In the ever-evolving landscape of asset inspection and maintenance, the ability to quickly gauge performance and make informed decisions is indispensable, as

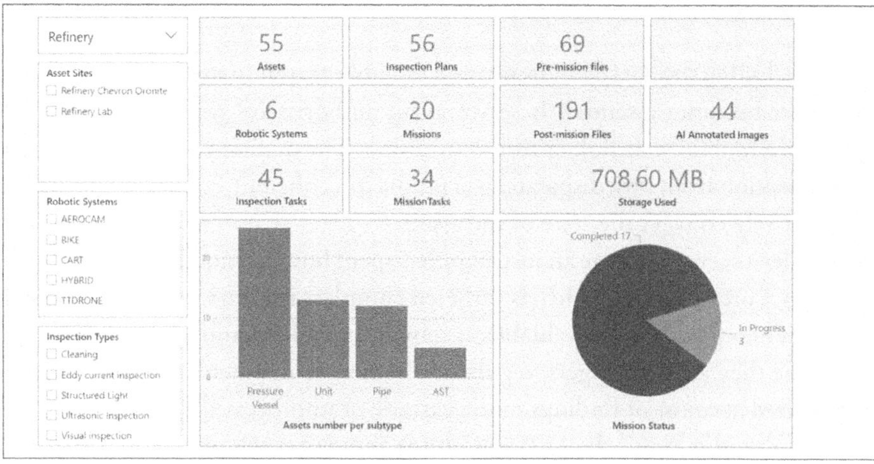

Figure 6.19. KPI Dashboard.

well as having a bird-eye view of the operations. Asset owners, inspection service providers, and stakeholders often have a multitude of questions: – how many assets have been digitized? Which ones have undergone inspection? – how many inspections have been completed? And so on. The KPI Dashboard addresses all these doubts.

Thanks to this dashboard (see Figure 6.19), users can dive deep into performance metrics, getting a holistic view of various KPIs concurrently. This overarching perspective ensures that stakeholders are always in the loop, understanding the nuances of their operations at a glance.

The dashboard provides users with advanced filters for a tailored experience, enabling individuals to sift through the data and pinpoint specific metrics that align with their criteria. This granularity ensures that users can zoom in on the information most pertinent to them, eliminating the noise and highlighting the signal.

The KPI Dashboard transforms complex datasets into structured, visually appealing graphs and charts, facilitating comprehension. Indeed, users can identify trends, note anomalies, and make comparisons with relative ease, ensuring that the decision-making process is both informed and reliable.

A tool's efficacy is often determined not just by its capabilities but also by its usability. The KPI Dashboard user-interface has been meticulously designed with this principle in mind. By prioritizing clarity, organization, and intuitive navigation, the dashboard ensures that users, regardless of their familiarity with the system, can access, interpret, and act on the data with minimal friction. The minimal design reduces the learning curve, enabling a broader spectrum of users to harness the power of the platform analytical capabilities.

6.4.6 Reporting Tools

Upon concluding the data analysis phase, a tangible output in the form of an inspection report becomes essential. It serves as an authoritative record, an actionable guide, and a tool for stakeholder communication. The question then becomes: how do we transition from raw, disparate data points to a coherent, structured document that effectively conveys the state of the asset?

Consider a scenario where an inspector has spent hours, if not days, meticulously examining a pressure vessel. He has gathered countless readings, images, and observations. While this data is invaluable, it remains nebulous until structured. Stakeholders, be they asset managers, regulatory bodies, or maintenance teams, require a synthesized account of findings, not a cascade of unprocessed data.

To cater to this need, the IVP introduces the report editor tool and the PDF report generation tool, both tailored to optimize the report generation process while maintaining the professional standards expected by asset owners.

The ability to generate a structured PDF report not only streamlines the inspection process but also enhances the usability and interpretability of the collected data. By amalgamating manually entered information with pre-existing asset data, the tool offers a holistic view of the inspection, ensuring that no detail, however minute, is overlooked. Furthermore, imagine the time-intensive process of crafting titles, creating tables, and ensuring a consistent layout for each report. The automation provided by the PDF report generation tool eradicates these repetitive tasks, enabling inspectors to concentrate on articulating their insights, findings, and recommendations.

In essence, the PDF Report Generation Tool transcends being a mere documentation tool. It embodies the bridge between raw data and actionable insights, ensuring that every stakeholder, regardless of their technical prowess, can glean the essence of the inspection findings. As we delve deeper into this section, we will uncover the myriad features that make this transformation possible.

6.4.6.1 Report editor

The Inspection Report Editor is meticulously designed to encapsulate the full spectrum of the inspection process, ensuring every facet is captured with precision:

- **Asset and Inspection Information:** At the heart of any inspection report lies foundational data about the asset itself and the particulars of the inspection undertaken. To streamline this, the system automates the incorporation of readily available data. By doing so, it mitigates the potential for human errors in data input, ensuring reports are both consistent and accurate.

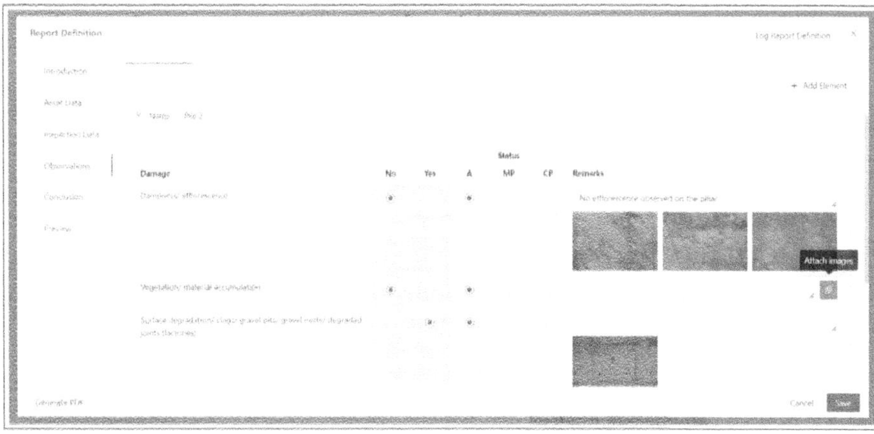

Figure 6.20. Observations table of the Report Editor tool.

- **Observation Information:** A crucial element of the inspection process is documenting findings. The editor offers a structured format for inspectors to document their findings in detail. The user can easily check the type of damage observed, indicate its severity and the urgency to take action, as well as attach images collected by the robotic payloads (see Figure 6.20). These features have been designed to facilitate the recording of all necessary information about the inspection, including visual evidence of any damage observed.
- **Preview:** Before finalizing the report, it is essential to assess its accuracy and completeness. The preview function serves this very purpose, allowing users to review the information in its entirety. This step acts as a final checkpoint, ensuring the generated report is both comprehensive and devoid of errors.

6.4.6.2 PDF report generation

Once all the data have been entered in the report editor tool, the last step is the transformation of data into a structured and accessible document, namely a PDF inspection report.

The portal employs a sophisticated mechanism when generating the PDF report. It amalgamates the manually entered information from the report editor with the pre-existing data related to the asset, site, and other pertinent details. This confluence of data is then systematically structured, with the portal autonomously crafting titles, tables, and more. The layout and design are derived from the chosen template, ensuring a consistent and professional presentation. This automation significantly reduces the administrative onus on inspectors, allowing them to channel their expertise into the actual inspection and analysis.

The evolution of the PDF generator is evident in its latest iteration, which comes equipped with a suite of features designed to augment the inspector documentation capabilities. The PDF Report Generation Tool can generate the following elements from the provided data:

- **Cover Page:** A consistent and branded cover page not only establishes authority and credibility but also sets the tone for the content that follows, making it immediately recognizable as a product of the PILOTING system. The portal automatically generates this cover page.
- **Table of contents:** Ensuring clarity in structure, the chapters and subchapters within the report are automatically recognized, numbered, and formatted. Clearness is at the core of the PDF generator. Furthermore, a structured table of contents is added at the beginning of the report, offering readers an immediate overview, and allowing for easy navigation of the document.
- **Tables:** Data presentation is crucial in an inspection report. The inclusion of tables allows for structured data presentation, ensuring that complex datasets are presented in an easily digestible format.
- **Observations tables:** Make it easier for the inspector to provide a comprehensive view of the asset condition by listing all observations made during the inspection. They help review and understand the findings of the inspection.
- **Image gallery:** Visuals provide undeniable evidence of the asset condition. Consider an inspection of a refinery tank; while descriptions can detail the corrosion or wear and tear, an image gallery offers a sequential visual documentation, showing the severity of rust patches or cracks. This aids stakeholders in understanding the urgency and extent of repairs required.
- **Cross-reference figures:** While discussing a defect on a bridge pillar in the text, cross-referencing allows readers to quickly jump to the exact image or graph that illustrates the described defect, ensuring comprehension. This feature ensures that any figure, table, or image can be easily referenced in the text, enhancing the coherence and interconnectedness of the content.
- **Highlighted text:** Among dense content, it is crucial that vital information stands out. Imagine reading about potential safety risks in an inspection report; highlighted text can immediately draw attention to areas with high accident potential, ensuring that safety measures are promptly taken. To achieve that, the PDF generator allows for text highlighting. Remarks can be highlighted in yellow, drawing attention without creating alarm. In contrast, urgent actions are highlighted in red, signaling immediate attention and priority.

In essence, the enhancements introduced to the PDF generator serve a dual purpose. First, they offer inspectors an enriched toolkit, simplifying the documentation process. Second, they enhance the quality of the resultant reports, making them more lucid and comprehensive. This dual benefit not only aids inspectors but also streamlines the decision-making process for stakeholders, ensuring that conclusions drawn from the reports are well-informed and backed by clear, detailed evidence.

6.4.6.3 Pre-defined templates

Consider an asset owner being presented with reports from various service providers. Each report, distinct in its layout, presentation, and structure, often leads to a muddled amalgamation of formats. One company's report is full of tables and charts, while another might be a verbose document with scattered images. Reading, comparing, and consolidating these myriad formats can be a challenging task for the asset owner. Such inconsistency not only impedes streamlined analysis but can also lead to crucial oversights.

To address this inconsistency, the PILOTING project introduced the predefined templates feature. Designed with the needs of the industry in mind, these templates ensure that regardless of the service provider, the report structure remains consistent. This uniformity allows asset owners to navigate through the report, effortlessly locate desired sections, and make informed decisions without getting entangled in format disparities.

Moreover, while these templates standardize the report layout, they also allow for customization to align with the branding of the asset owner company. This ensures that while the report's structural integrity remains intact, its aesthetics can mirror the company visual identity, adding a touch of personalization. Such a blend of standardization and customization strikes the right balance, ensuring that the report is both recognizable and uniquely tailored. An example is shown in Figure 6.21.

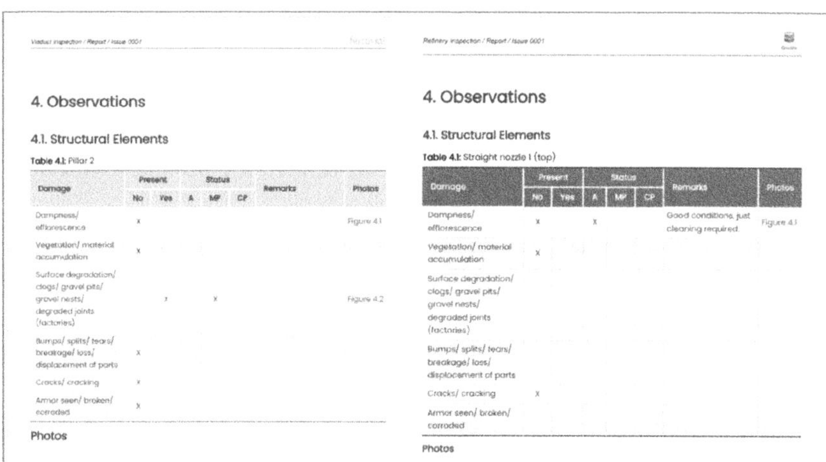

Viaduct inspection / Report / Issue 0001

4. Observations

4.1. Structural Elements

Table 4.1: Pillar 2

Damage	Present		Status			Remarks	Photos
	No	Yes	A	MP	CP		
Dampness/ efflorescence	x						Figure 4.1
Vegetation/ material accumulation	x						
Surface degradation/ clogs/ gravel pits/ gravel nests/ degraded joints (factories)		x		x			Figure 4.2
Bumps/ splits/ tears/ breakage/ loss/ displacement of parts	x						
Cracks/ cracking	x						
Armor seen/ broken/ corroded	x						

Photos

Refinery inspection / Report / Issue 0001

4. Observations

4.1. Structural Elements

Table 4.1: Straight nozzle 1 (top)

Damage	Present		Status			Remarks	Photos
	No	Yes	A	MP	CP		
Dampness/ efflorescence	x		x			Good conditions, just cleaning required.	Figure 4.1
Vegetation/ material accumulation	x						
Surface degradation/ clogs/ gravel pits/ gravel nests/ degraded joints (factories)							
Bumps/ splits/ tears/ breakage/ loss/ displacement of parts							
Cracks/ cracking	x						
Armor seen/ broken/ corroded							

Photos

Figure 6.21. Illustration of the transformation from the report editor (see Figure 6.20) to the final PDF format, showcasing two distinct customized templates for Ferrovial (viaduct use case) and Chevron (refinery use case). The main layout is consistent between different use cases.

For service providers, the benefits are varied. Firstly, the use of a predefined template eliminates the arduous task of designing and/or adapting a report layout for every inspection, thereby saving time, and ensuring productivity. Additionally, knowing that their report aligns with an industry-standard format can boost their confidence in the presentation and deliverability of their findings.

In essence, the introduction of predefined templates is a strategic move towards fostering cohesion in the reporting process. Addressing the real-life challenges faced by both asset owners and service providers, it ensures clarity, consistency, and a touch of customization, setting conformed with asset inspection reporting standards.

6.5 Conclusion

In the evolving landscape of asset inspection, the integration of digital solutions has emerged as a game-changer. The PILOTING project, as we have explored, serves as a paragon in this digital evolution, demonstrating innovative methodologies with practical applications. The project not only challenges traditional inspection paradigms but also offers solutions that are both robust and user-friendly.

One of the standout features of this initiative is the Intelligence and Visualization Portal (IVP). The IVP is more than just a digital interface; it is an end-to-end solution that spans the entire inspection journey – from meticulous planning to detailed reporting. By leveraging 3D digital twins, it facilitates a data-driven approach that simplifies complex inspection tasks while ensuring precision and reliability.

An integral aspect of the IVP prowess lies in its ability to embrace multiple robotic platforms. In a world teeming with diverse robotic solutions, each tailored for specific tasks, the portal's capability to integrate and harmonize data from various robotic systems is truly commendable. Whether it is the intricate details of a pressure vessel or the expansive architecture of a tunnel, the portal ensures that the right robotic system is employed for the job.

Furthermore, the emphasis on context cannot be overstated. In our exploration, we saw how features like trajectory mapping, image contextualization, and interactive UT measurement in a 3D environment provide invaluable context, transforming isolated data into a coherent visual narrative that helps inspectors make informed decisions.

The suite of tools available for image annotation and AI-enhanced analysis further accentuates the portal's commitment to accuracy and efficiency. By allowing inspectors to annotate images, be it through bounding boxes or arrows with text, and then supercharging this process with AI-driven insights, the portal ensures that no anomaly goes unnoticed.

But what truly stands out is the portal's emphasis on effective communication. The meticulous process of inspection culminates in the generation of a comprehensive report, a document that serves multiple stakeholders. The PDF Report Generation Tool, as we have discussed, transforms raw data into a structured narrative, ensuring that the insights gleaned from the inspection are effectively communicated, be it to an asset manager, a regulatory body, or a maintenance team.

In closing, the PILOTING project showcases the transformative potential of digital solutions in the realm of inspection and maintenance. Integrating state-of-the-art technologies with user-centric designs, the project not only elevates the inspection process but also sets a benchmark for future innovations in this domain.

As asset infrastructures continue to age and the demand for streamlined inspection solutions grows, initiatives like PILOTING will undoubtedly play a pivotal role in shaping the future of asset maintenance and management.

Acknowledgements

The authors of this chapter are grateful to the PILOTING technical partners and end-users for providing continuous feed-back during the development and integration of complete platform. The authors would like to also thank Maarit Sandelin for her contribution.

This project has received funding from the European Union Horizon 2020 Research and Innovation Programme under grant agreement No. 871542.

References

[1] The PILOTING Project, https://piloting-project.eu/.

[2] Zenodo community, https://zenodo.org/communities/piloting/?page=1&size=20.

DOI: 10.1561/9781638282839.ch7

Chapter 7

Autonomous Navigation for Inspection and Maintenance Ground Robotics

By Peter Ørnulf Ivarsen, Julian Keller, Jan Sramota, Victor Reijgwart, Martin Eek Gerhardsen, Cesar Cadena, Henrik Lundkvist and Richard J.D. Moore

Autonomous inspection and maintenance operations have the potential to reduce the cost, increase the quality and enable new functionality. Robust and accurate navigation for ground vehicles is a key enabler for safe operations and reliable location tagging of inspection data. A major challenge in the refinery and tunnel environments that we target in PILOTING is the lack of reliable GNSS coverage, which calls for other methods for localization. This has been addressed by developing SLAM solutions with sensors and software adapted to the specific environments. These solutions enables use cases in autonomous maintenance and inspection operations, which has been validated in refineries and tunnels respectively. This chapter describes the autonomous navigation for ground robots developed in PILOTING, and their performance.

7.1 Introduction

Autonomous navigation is one of the most important functionalities in order to achieve an automatic inspection using robotic systems, even more important in environments without GNSS coverage (e.g., inside tunnels) or with degraded GNSS due to proximity to obstacles (like in refineries). Autonomous navigation algorithms for ground robots in indoor, static and well-structured environments have been developed to the state of consumer-grade deployment. An important aspect for this success is the reached maturity of simultaneous localization and mapping (SLAM) systems over the last decade. This progress has led to reliable systems based on single modalities, e.g., LiDAR or camera data [10].

For autonomous mobile robots, accurate navigation stands as a cornerstone for the efficient execution of inspection and maintenance tasks. The challenges inherent in navigating within diverse and complex environments, where reliance on GNSS is often untenable, have driven the development of innovative solutions. This is particularly evident in PILOTING scenarios. Here, the ability to seamlessly integrate both mapping and localization functionalities becomes pivotal.

Localization, precisely determining the robot's position within its environment, serves a dual purpose in these endeavors. It not only facilitates the autonomous navigation of vehicles but also plays a pivotal role in tagging inspection data with accurate spatial information. This seemingly straightforward task of location tagging assumes profound significance, enabling the inspection platform to preserve vital information over time and observe nuanced changes in the condition of the environment.

For autonomous robots to carry out maintenance tasks effectively, the need for adept object manipulation methods cannot be overstated. Mapping and localization lay the foundation for successful navigation, and effective manipulation capabilities need to be developed to follow inspection and maintenance plans autonomously.

In this chapter, we delve into autonomous navigation, exploring its importance in scenarios where GNSS coverage is unreliable or nonexistent. We will present the localization, mapping, and their importance, together with path planning, for autonomous navigation, and as enablers of use cases: (i) location tagging for long-term observations, and (ii) object manipulation in the pursuit of autonomous maintenance tasks.

7.2 Localization Algorithms

To provide some context, we first introduce localization methods that rely on external systems that provide reference signals for the positioning. Then we focus on SLAM solutions, which are necessary in the scenarios considered in PILOTING.

7.2.1 GNSS and Radio-Based Localization

Global Navigation Satellite System (GNSS) localization relies on a network of satellites and ground-based receivers, to provide precise and reliable positioning, velocity, and timing information for a wide range of applications. There are multiple satellite constellations deployed, including the United States' GPS, Russia's GLONASS, Europe's Galileo, and China's BeiDou. By measuring the time it takes for signals to travel from satellites to receivers the position of the receivers can be triangulated.

GNSS localization faces challenges such as errors caused by signal distortions, atmospheric interference, and other factors, ultimately providing accurate position and timing information. Various mitigation techniques, such as antenna design, signal filtering, and advanced algorithms, have been developed to combat these challenges and improve localization accuracy. Two prominent techniques are correction signals and Real-Time Kinematic (RTK) positioning.

Correction signals provide additional information to the GNSS receiver, enabling it to compensate for various error sources. There are different types of correction signals available: Satellite-Based Augmentation Systems (SBAS) to broadcast correction data to GNSS receivers. These correction signals account for ionospheric delays, clock errors, and other sources of error, significantly improving positioning accuracy. Differential GNSS (DGNSS) involves using a reference station with known coordinates to calculate the errors in GNSS measurements caused by factors like atmospheric delays and satellite clock errors. The reference station transmits correction data to nearby receivers, allowing them to correct their position calculations accordingly.

Real-Time Kinematic (RTK) positioning is a highly precise GNSS technique that leverages carrier phase measurements to achieve centimeter-level accuracy in real-time. RTK requires a base station located at a known position, which measures carrier phase ambiguities and transmits them to the receiver in real-time. By resolving these ambiguities, the receiver can achieve highly accurate positioning relative to the base station.

7.2.1.1 Radio-based localization in GNSS denied environments

While GNSS localization has revolutionized positioning and navigation, there are scenarios where line-of-sight between GNSS satellites and receivers is obstructed. Radio-based localization techniques are similar to GNSS in that they use radio signals to estimate the position, however these radio signals are sent between terrestrial transmitters and receivers and can therefore use transmissions both to and from the device that shall be positioned for the estimation. Bluetooth beacons and Ultra-Wideband (UWB) technologies are often utilized for localization purposes. These techniques are especially useful in indoor environments where the coverage area is limited and it is feasible to install transmitters that provide good coverage. Alternatively, communication systems such as mobile networks or Wi-Fi can also be used for positioning. In particular, the positioning accuracy provided by 5G mobile networks is improving with the evolution of the technology through new releases of the standards, but also due to the increased bandwidth availability and the denser deployment of antennas.

7.2.2 Sensor Fusion Based Localization

While radio-based localization techniques have the advantage that the cost of the equipment on the device is low, they do require infrastructure in the form of radio transmitters to be deployed. Positioning methods are based solely on sensors on the device that shall be positioned are therefore often more attractive. To overcome limitations of individual sensor types, sensor fusion techniques amalgamate data from multiple sensors to estimate position and map the environment. In addition to cameras and inertial sensors, this can include lidar and radar for distance and velocity estimation, magnetometers that leverage the Earth's magnetic field to estimate orientation, and barometers which use the air pressure to estimate altitude. In the rest of the chapter we will focus on localization techniques that use multiple sensors.

7.2.2.1 Simultaneous Localization and Mapping (SLAM)

Simultaneous Localization and Mapping (SLAM) is a technique that involves estimating a robot's state, typically described by its position and orientation, while simultaneously constructing a map of its environment based on sensor input. In more complex scenarios, the robot's state can include additional parameters like velocity, sensor biases, and calibration data. The best-performing SLAM algorithms use pose-graph systems for efficient data association and map optimization [11, 12]. However, challenges arise in dynamic and challenging conditions, such as outdoor and industrial settings, leading to incorrect data associations. Robust estimators [9] and geometric verification are essential to address these issues, with loop closure and global localization techniques being vital for map consistency. Vision-based systems encounter challenges like viewpoint changes and poor-textured areas [10, 13], while geometry-based systems struggle with data ambiguities and memory-intensive representations [16]. These perceptual challenges can be tackled through unified representations [15] or a combination of spatial and appearance data fusion [17].

The perceptual challenges discussed above manifest prominently in practical applications, particularly in the context of PILOTING scenarios, such as those encountered in complex environments like refineries and tunnels. Refineries pose unique challenges due to the presence of self-similar features and surface materials with high reflectivity. The structural components often comprise steel pipes, grids, and wireframes, making it challenging for conventional robotic perception methods to detect thin structures and essential details. Consequently, the incorporation of complementary sensor modalities becomes imperative for successful navigation, inspection, and manipulation tasks within the refinery environment. Similarly, tunnels pose their own set of difficulties, characterized by poor lighting conditions and a high degree of self-similarity, rendering SLAM-based localization particularly challenging in this context.

7.3 Navigation Algorithms

Navigation algorithms play a vital role in guiding vehicles, robots, and other autonomous systems through their environments. Through path planning, obstacle avoidance, and trajectory optimization, these algorithms enable vehicles, robots, and other autonomous systems to navigate complex and dynamic surroundings.

The first part of the navigation is to plan a path based on input from a user interface or any other tool that provides the waypoints that shall be visited, or area that shall be covered by the vehicle. Path planning algorithms determine the optimal path from a starting point to a desired destination while considering various constraints and objectives. These algorithms take into account factors such as known obstacles, vehicle dynamics and environmental conditions. A typical solution is the A* search algorithm, which uses a greedy search algorithm to find a low cost path through a graph based on some cost function [1].

To handle dynamic obstacles that are unknown at the time of path planning, the vehicles rely on sensors such as security lidars for detection. To allow effective autonomous operation the UGV needs to continue operating around the obstacles rather than just stopping. Obstacle avoidance algorithms enable autonomous systems to navigate around obstacles encountered during their path that were not known during the initial path planning. The theme of motion planning and control will be further presented on the example of refinery navigation in the next sections.

7.4 Localization and Mapping of Ground Robots in PILOTING

In this section perception and navigation solutions for ground robots developed in the PILOTING project for tunnel and refinery scenarios are described including both hardware and software. Common for these navigation solutions are that they are based on onboard sensors and onboard processing that allow the robots to operate autonomously. This reduces the need for external infrastructure and communication. However, it increases the requirement on computing power and makes the selection of sensors critical to achieve the navigation requirements with a reasonable energy consumption and component cost.

The inspection missions are provided from the PILOTING I&M platform, where they have been planned using a 3D map. The map may either be created from existing models, or they can be generated by the PILOTING inspection robots. It is important that the 3D maps shall be consistent over time, so that it is possible to localize and monitor possible changes of the state of defects and assets of interest.

7.4.1 Refinery Navigation

The navigation in a refinery is challenging with large structures that obstruct GNSS signals, therefore a SLAM solution is needed. The UGV in the refinery is a skid-steered 4-wheeled robot equipped with a 3D LiDAR, visual-inertial tracking camera, wheel encoders, and onboard computing. The navigation stack is tailored to large industrial environments such as refineries, which motivates the assumptions that the environment does not change often and is only shared with cooperative agents and humans.

7.4.1.1 Localization and mapping

A prior map of the environment can help robot localization and navigation, especially if this map can be assumed to be mostly static over time. This is the case for typical industrial environments, which is why this approach can be chosen confidently in such scenarios. In contrast to more dynamic approaches relying on other sensor modalities, such as GPS, this approach also works, for example, in environments with degraded sensor reception due to metallic structures. We have split the localization and mapping pipeline of the robot into two modes. In the first mode, the robot performs LiDAR-inertial Simultaneous Localization and Mapping. This mode is typically used while driving the robot around the facilities under teleoperation. Alternatively, a human operator can scan the plant by hand using a single LiDAR-inertial sensor, such as an Ouster OS1, and a laptop or lightweight computer. Given recent advances in LiDAR sensors and LiDAR-inertial odometry, the global maps resulting from FastLIO2 exceeded the necessary quality and accuracy in all performed experiments. Note that FastLIO2 does not explicitly detect global loop closures and apply corresponding global map deformations. Although it turned out not to be necessary in the environments tested thus far, a pipeline was developed to incorporate global loop closure corrections in Open3D SLAM [18] if needed. The resulting precise, dense 3D point cloud map of the environment

Figure 7.1. Image of a pre-built map for missions in a refinery environment, recorded using a single Ouster OS1-64 LiDAR sensor and its integrated IMU.

can then be used to plan subsequent inspection and manipulation missions. An example of a 3D point cloud map can be seen in Figure 7.1. The main difference with respect to previous solutions is that the mapping module no longer requires a visual-inertial input. Removing reliance on time synchronized cameras reduces the need for specialized hardware and intrinsic and extrinsic calibration. Furthermore, by virtue of being active sensors, LiDARs are insensitive to illumination changes and most weather conditions.

7.4.1.2 Motion planning and control

Motion planning is crucial for the robot to navigate to a given goal while avoiding static and dynamic obstacles along the way. The motion planning is typically divided into a global and a local planning phase. Defining the inspection locations of interest and the order in which they should be visited is generally done by means of a coordination portal software, which enables the operator to instruct the global motion planning for the robot. Using the pre-built map as a basis, the global navigation can be done efficiently by means of an industry-standard RRT* planner. It calculates the shortest path from the robot's current position to the next mission waypoint while considering the robot's footprint and obstacles in the pre-built map. The global planner assumes that the environment is static, but it can be helpful if it supports traversability annotations that can for example be used to mark dangerous areas, invisible obstacles or areas where the ground is not safe to drive on as untraversable.

The local planner then translates the global plan into safe local trajectories that: (1) avoid collisions with newly encountered obstacles in addition to the obstacles of the pre-built map, (2) stay close to the global path while allowing for small detours (e.g. around displaced equipment or dynamic obstacles including humans), and (3) ensures that only kino-dynamically feasible trajectories are sent to the controller. These requirements can for example be met by using an online optimal local time elastic band (TEB) planner. Our experiments in refinery environments have shown that the TEB planner effectively and safely follows coarse paths from the global planner, while avoiding dynamic, local obstacles and effectively negotiating small differences between the pre-built map and the robot's real surroundings. Furthermore, the experiments have shown that by explicitly considering the robot's specific constraints (e. g. ackermann steering or skid-steering), the TEB planner generates trajectories that are feasible and can accurately be followed. This can be done by sending them to the robot platform's low level controller. This direct control approach has proven efficient, robust and easy to maintain in practice, since no intermediate controller (such as an MPC) that would add additional complexity is required.

7.4.1.3 Use case: Autonomous valve manipulation

The autonomous navigation solution addresses the general issue of building a persistent representation of the environment and localizing the robot within. For the refinery scenario a UGV has been equipped with grippers to have the means to execute maintenance tasks. In this section the solution for perception and control is described. The UGV with gripper can be seen in Figure 7.2. The task that is being addressed is manipulation of valves, which is a useful capability that is not supported by currently commercially available robots for refineries. Since the handwheels on the valves are all designed for human workers, the ground robot needs to be designed taking into account the physical constraints that this implies, in order to not damage the assets of the refinery, or the robot itself.

We focus on the autonomous operation of an UGV within a (quasi-static) typical industrial refinery environment. The foundation for precise dexterous manipulation tasks lies in the prior knowledge of the locations and types of assets, such as valves, present within the environment. This information can be extracted from refinery documentation or digital building information models, seamlessly incorporated into a digital twin of the refinery generated through the mapping process outlined in Section 7.4.1.1. The minimal requirement for successful valve manipulation is the knowledge of their locations, allowing for a low-entry barrier and cost-effective adoption of the manipulation pipeline at new plants. With the asset locations determined, the operator specifies the exact properties of the manipulation task, particularly during inspection mission planning, where inspection waypoints are set for each valve, including turning angles and directions.

Figure 7.2. Image of the UGV equipped with a manipulator. The inspection plan provides the location of the asset in the pre-built map, then the UGV localizes within the same and navigates towards the asset's location and proceeds with the manipulation.

Once all parameters are set, the manipulation mission can be autonomously executed by the robot, reducing the need for operator intervention and thereby enhancing cost-efficiency. Throughout the mission, relevant data, including robot telemetry, manipulator status, joint torques, and end-effector forces, are captured, along with warnings and errors. This data serves as a valuable resource for improving autonomous missions over time and for early detection and resolution of errors or deviations from the digital model, enabling operators to focus on high-level tasks and reducing long-term manual efforts. In the context of valve manipulation, the process involves perceiving the 3D pose of the valve, deploying a dynamic grasp strategy based on manipulator and asset positions, generating manipulation plans with motion primitives, and tracking end-effector trajectories using a compliant cartesian control scheme with feedforward gravity and Coriolis compensation, ensuring robust performance even in the presence of minor detection or alignment issues.

7.4.2 Tunnel Navigation

The tunnel inspection will be done in two different phases, a general inspection and a local inspection. In the general inspection, the UGV will drive through the tunnel and collect inspection data using its onboard inspection sensors. From this data it will be determined where there is a need for more detailed inspection of defects and other points of interest. The UGV will then return to those places and carry out a detailed inspection where inspection data can be collected at close distance, for example using a drone and an arm mounted camera.

Compared with the refinery navigation, the tunnel navigation has simpler path planning since the UGV only need to follow the tunnel wall at a constant distance. However, the geometry of the tunnel is more monotone, which makes the localization more challenging.

7.4.2.1 Challenges for localization and mapping in tunnels

The main challenges of delivering well-functioning localisation in the case of tunnels are connected to the particular conditions we find in such environments:

1. Geometry is not locally unique, hence traditional LiDAR based SLAM solutions will not be able to assess ego-motion uniquely in any straight part of a tunnel.
2. The environment contains repeated structures such as signs, lamps and road markings. This will make global place recognition vulnerable to proposing erroneous data association and loop closures.
3. In trafficked tunnels, a large part of the environment can consist of dynamic objects such as vehicles or cyclists. If the SLAM system bases its estimated ego-motion on these objects instead of the static environment, errors will occur and accumulate.
4. Light is usually limited and varying.

SLAM solutions for tunnel inspection have been evaluated by Filip *et al.* in [2], where several state-of-the-art LiDAR-based SLAM solutions were tested and it was found that the positional root mean square error (RMSE) for all methods to be several meters on a 40 m trajectory. For the inspection use-cases this error is too large, in PILOTING the goal is to achieve errors in the order of a decimeter. To improve the accuracy and reliability of the lidar based SLAM, additional sensors are needed.

7.4.2.2 PILOTING solution

The SLAM solution developed in PILOTING is based on LOAM – Lidar Odometry and Mapping in real-time [3]. This combines an odometry thread that computes

Figure 7.3. The navigation payload.

the motion of the LiDAR between sweeps with a mapping thread that incrementally builds a map from the point cloud and computes the pose of the LiDAR in the map. To make the solution robust visual input from a camera adds constraints to the state estimates, as well as optional wheel odometry.

The UGV is equipped with a navigation payload with three sensors: a 3D LiDAR, a camera and an IMU. In addition, it has an onboard computer that performs all the computations of the SLAM algorithm. The navigation payload can be seen in Figure 7.3. The algorithm requires that data from all three sensors must be precisely time-synchronized. The LiDAR and the onboard computer supports precision time protocol (PTP) in hardware which achieves a sub microsecond synchronization. Camera frame acquisition is governed by a trigger signal from the LiDAR when the LiDAR beam is at 0 degrees, i.e. aligned with the camera direction. This way data from all sensors are collected and time-aligned in the onboard computer.

A block diagram of the SLAM operation can be seen in Figure 7.4. The LiDAR cloud is undistorted by integrating up the IMU's gyro since the last scan and thus finding the UGV's rotation during the scan period. This compensates for the movement of the UGV during the acquisition of the LiDAR scan. Then surface features are extracted from the LiDAR scan. The LiDAR based estimate of the robot pose in a tunnel is well constrained in five out of six degrees of freedom, namely the rotational components and the translation in the height and width dimensions of the tunnel. However, the translation along the traveling axis of the tunnel will drift as the geometry of each scan along this axis will be more or less identical.

To address the drift in the state estimation the camera is used to add nongeometric features that constrain the optimization also along the degenerate axis. ORB features [4] and ArUco features [5] are extracted from the camera image in parallel with the LiDAR processing. ORB features are extracted per scan and added as additional locally unique constraints in both the scan matching performed in the odometry thread and in the mapping thread. These might be extracted from light

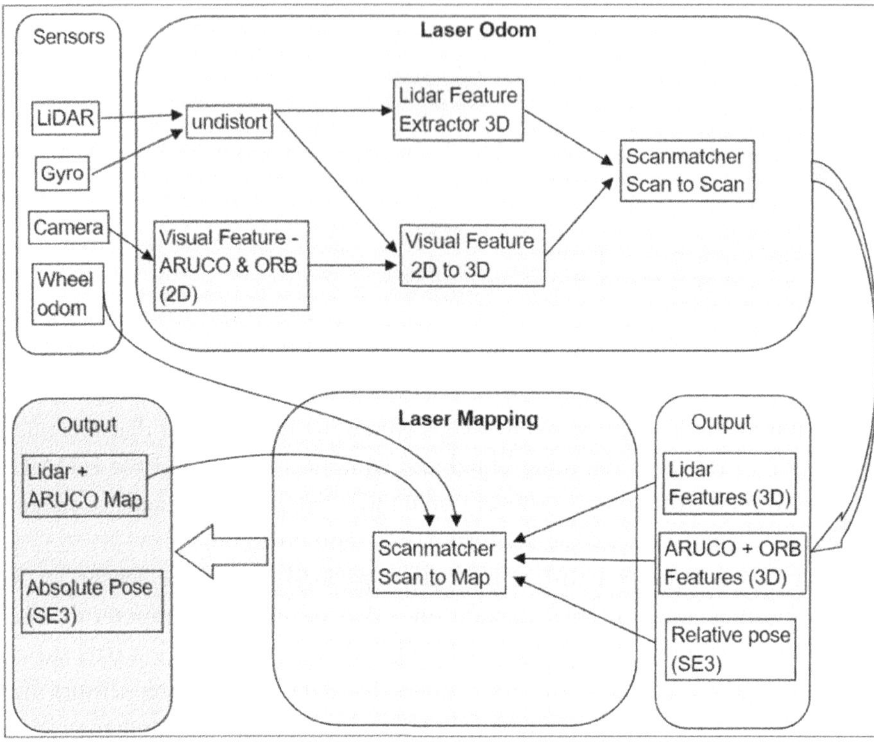

Figure 7.4. A schematic of the localization and mapping algorithm. Ceres is the optimization framework that is used in the ICP scan matching.

fixtures, road markings, signs, and markings on the tunnel wall. ArUco features are extracted from QR codes that have been placed at known positions and are added to the scan matching in the mapping thread. These are considered globally unique and can thus be used to correct for drift in the motion estimate when the robot is visiting a place it has seen before, or when localizing in an existing map.

To determine the 3D coordinates of the visual features they are projected into 3D by sampling the LiDAR cloud for depths, as illustrated in Figure 7.5. Following this, scan-to-scan matching is performed in the odometry thread, matching surface features and ORB features from the previous and the current scan to estimate the relative motion between scans.

Motion estimates from the odometry thread as well as surface, ORB and ArUco features are passed on to the mapping thread. Here, scan-to-map matching is performed to align the current LiDAR data to the map.

After scan-to-map matching is successful, all incoming point features are added to the map.

Wheel odometry from the UGV can instead be added as an optional constraint in the scan matching performed in the mapping thread. The wheel odometry can

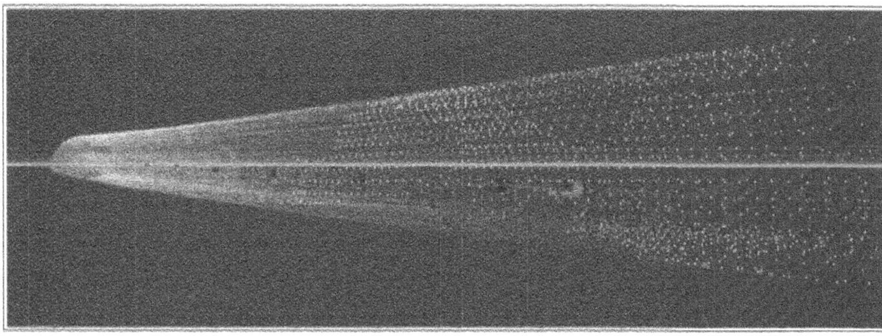

Figure 7.5. UGV pose (red arrow), surface features (pink dots), ORB features (smaller red/blue squares) and ArUco features (large blue cubes) projected into the point cloud.

be considered as a redundancy in the system and will contribute to constraining the translation along the traveling distance of the tunnel in case a sufficiently large set of visual features is not available.

7.4.2.3 Map creation

In most cases a map will be available beforehand from an inspection and maintenance platform such as PILOTING. This allows repeatable observations of faults in a tunnel over time in a common reference frame. However, the system shall also handle the case were a new 3D map must be created. Then it is important to that the map includes sufficient information to correct for drift, contains as little noise as possible, and is as accurate as possible.

To minimize drift in a tunnel that can be very self-similar for both LiDAR and camera sensors, a solution is to mount explicit landmarks such as ArUco markers prior to mapping. These will be registered by the SLAM method to the map, which will thus consist of a dense point-cloud representing the geometry of the tunnel, in addition to a set of points per observed tag with a unique identifier.

Noise may be introduced in the map by dynamic objects such as cars. When the system then reuses the map, these dynamic elements can align with current dynamic elements and introduce errors in the estimated motion. To avoid this issue a postprocessing step is applied to remove such dynamic features stored in the map. For example, the Removert algorithm [7] can be used to clean the map in case of tunnels with traffic, removing most dynamic features.

To improve the quality of the map, a solution that is used in PILOTING is to utilize a total station to constrain the UGV position during offline map building. This requires some setup and is in practice done offline both because the total station lacks long range communications with the UGV and due to that coordinate frames and clocks between the UGV and the total station need aligning. This workflow requires a larger setup, but allow for a theoretical sub-centimeter accuracy when building a map of a new site. It is worth noting that even though the total station is a valuable constraint for SLAM, it is not suitable for deployment on long range inspections due to the line of sight requirement.

7.4.2.4 Performance evaluation

In the evaluation of the system, the additional constraint from the total station is only used to create the map used for localisation. During the evaluation of the localisation performance, this map is used, but only LiDAR, camera, and IMU data is used for localisation – the total station is only used for ground truth. Our method is evaluated both when localising in an unknown environment and when we are localising in an existing map.

The system has been validated in a 175 m long former railway tunnel outside the city of Coripe, Spain. Accompanying these recordings we have position measurements, as reported by a Leica robotic total station tracking a prism mounted on the UGV. The total station reports position with sub-centimeter precision at 10 Hz, and can thus be considered as a ground truth with which to evaluate the system.

We have three data-sets, the first set (A) is a round-trip in the tunnel at inspection speed, about 1 m/s, with only the tunnel lights used as illumination. The second data-set (B) is recorded under the same conditions but with a speed of about 4 m/s. And the third data-set (C) contains a single forward pass with a similar speed as (B) and with the vehicle's headlights switched on.

We have also benchmarked two existing SLAM systems against ours, the LiDAR inertial SLAM system named LIO-SAM [8] and a LiDAR visual inertial SLAM system, LVI-SAM [9]. We have made a reasonable attempt to achieve good performance from the two algorithms, but further tuning could likely be made to improve results.

As an evaluation framework we have used TUMs evo package for comparison of trajectory output of SLAM algorithms [6]. For the reported RMSE we have aligned the estimated trajectory with the ground truth. As seen in Table 7.1 both LIO-SAM and LVI-SAM have large errors, as motion estimation deteriorates when

Table 7.1. Errors for different methods after full trajectory for the three data sets.

Method	A – RMSE (m)	B – RMSE (m)	C – RMSE (m)
LIO-SAM	52.79	51.87	52.49
LVI-SAM	44.48	28.84	45.25
PILOTING	0.61	0.31	0.40
PILOTING (with map)	0.09	0.16	0.11

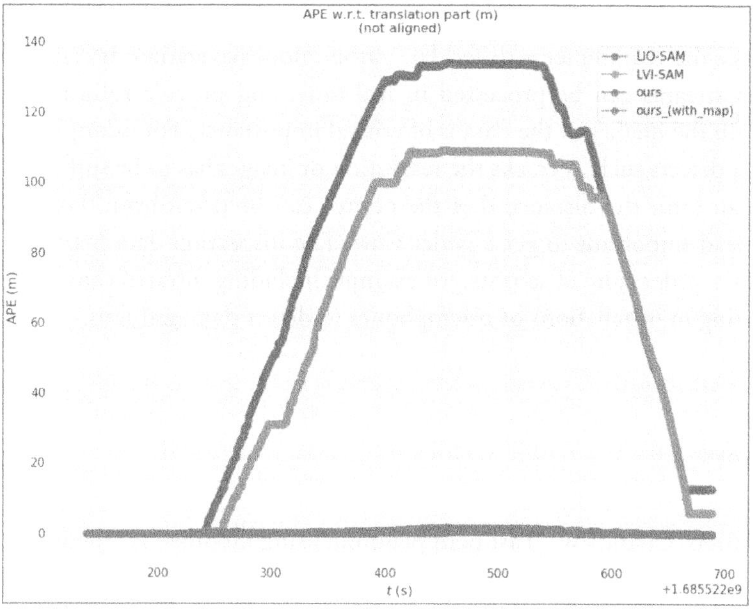

Figure 7.6. Comparison of the absolute positional error of the estimated trajectories based on data-set A, a pass back and forth through the 175 m long tunnel at inspection speed.

the LiDAR no longer observes structures from outside the tunnel. This is likely due to the lack of strong structural features inside the tunnel. The proposed PILOTING system running without a map achieves an RMSE well below a meter for the three trajectories, and below 20 centimeter when allowed to correct for drift using a pre-built map.

For further comparison one can observe the absolute positional error (APE) over time plotted in Figure 7.6. From this we can see that the LVI-SAM is able to correctly estimate the position of the UGV for a longer time then then LIO-SAM and also that the error at the end of the recording, when the UGV is back to the start position is smaller, indicating that the addition of a camera adds to the accuracy. As for our suggested approach we see that running the system without a map result in a larger error, and that the absolute error is largest at the far end of the tunnel.

7.4.2.5 Use case: Autonomous inspection

The autonomous navigation in tunnels is the basis for an autonomous inspection solution. An UGV is provided with a mission to inspect an area or points of interest in a tunnel to provide data that can be used for both automatic and manual data analysis. The accurate localization is a key to make the inspection efficient, since the re-localization of previously detected damages can be made quickly.

DOI: 10.1561/9781638282839.ch8

Chapter 8

Application of Intelligent Aerial Robots to the Inspection and Maintenance of Electrical Power Lines

By Anibal Ollero, Alejandro Suarez, Juan Manuel Marredo, Giovanni Cioffi, Robert Pìnièka, Goran Vasiljević, Viet Duong Hoang, Michele Marolla, Jiaxu Xing, Martin Saska, Stjpan Bogdan, Emad Ebeid, Fabio Ruggiero, J. Ramiro Martínez-de Dios, Davide Scaramuzza, Vincenzo Lippiello and Antidio Viguria

This Chapter deals with applying several technologies developed in the H2020 project AERIAL-CORE (AERIAL COgnitive integrated multi-task Robotic system with Extended operation range and safety) to inspect and maintain electrical power lines. The Chapter includes the following methods and technologies developed in AERIAL-CORE: application of perception-aware model predictive control for the aerial tracking of electrical power lines without prior information about the power line infrastructure; long-range powerline and vegetation mapping by using a 3D solid state LiDAR and a RGB camera for user visualization; aerial manipulators for the installation of bird-diverters and a recharging station, by including a high payload light arm, a linear actuator platform, a dual arm platform, a large general-purpose 45 kg take-off weigh manipulation platform and magnetic gripper to battery charging from the power line; human-machine interfaces; and a formation of aerial co-workers to monitor the safety of a human worker performing maintenance activities on the powerline. Experiments with real electrical power lines are included in the paper.

8.1 Introduction

It is well known that Unmanned Aerial Systems and Aerial Robotics have the potential to bring a revolutionary transformation in the industrial inspection market [1] by increasing productivity, reducing inspection time, improving data quality, and eliminating the risks for human operators. They can be used to regularly inspect the power line infrastructure to prevent power outages and natural disasters [2].

Aerial Robotics emerged in the nineties of the last century, fueled by the development of Unmanned Aerial Vehicles (UAVs), either fixed-wing or rotary-wing platforms, providing basic onboard autonomous functionalities. In the last decade, aerial robotics has experienced a very important growth with significant perception, reactivity and planning capabilities. Autonomous detection, tracking, simultaneous localization and mapping and other functionalities have been developed.

Aerial robotic manipulation also emerged by the beginning of the second decade of the last century. The FP7 ARCAS (Aerial Robotics Cooperative Assembly System) project (2011–2016) played an important role in developing general control, perception and planning functionalities. Later, the H2020 AEROARMS (AErial RObotic system integrating multiple ARMS and advanced manipulation capabilities for inspection and maintenance) project (2015–2019) further developed aerial robotic manipulation and applied it to inspection and maintenance. Surveys in Aerial Robotic Manipulation can be found in [3, 4].

However, by 2019, inspection and maintenance based on aerial robotics, and particularly aerial robotic manipulation, was mainly constrained to local interventions. Also, aerial manipulation was limited to contact inspection without the applications of the forces required to perform many maintenance activities. Moreover, the safety of aerial robots in inspection and maintenance activities and the collaboration with human workers were not considered.

Furthermore, at that time learning and other Artificial Intelligence techniques offered many new possibilities for detection, tracking, recognition and real-time decision and control.

The main objective of AERIAL-CORE is the development of innovative aerial robotics technologies resulting from the application of Artificial Intelligence. Notably, we are looking to extend the operational range of aerial robots, improve the performance of aerial manipulators, and increase safety in the interaction with people for applications such as the inspection and maintenance of large infrastructures. Thus, AERIAL-CORE targets the Research and Innovation in the following topics:

(1) Long-range, accurate inspection of the infrastructure.
(2) Maintenance activities based on aerial manipulation.
(3) Aerial co-working helping human workers in inspection and maintenance tasks.

AERIAL-CORE includes the application of the above methods and technologies to the inspection and maintenance of Electrical Power Lines, which is very challenging application with a strong impact.

In this Chapter, we present several AERIAL-CORE results in the above topics. Thus, the following section describes intelligent systems for long-range inspection. The third section deals with aerial robotic manipulation for maintenance and includes installing battery recharging stations on the line. The fourth section focuses on human-machine interfaces, and the fifth section on multiple-UAV for monitoring the safety of human workers. Finally, the sixth section is devoted to the conclusions and future work.

It should be noted that not all the AERIAL-CORE technologies are included in this Chapter. In particular, UAV morphing and bioinspired technologies used to design and develop new platforms that can combine long-range and local inspections are not considered.

8.2 Intelligent Systems for Long-Range Inspection

Intelligent systems for long-range inspection in AERIAL-CORE include a variety of technologies such as sensor data fusion, learning techniques, line tracking, mapping and planning of multiple aerial systems. In this section, only the line tracking and mapping are included.

8.2.1 Perception Aware Model Predictive Control for Inspection of Power Lines

We propose [5] a vision-based, tightly-coupled perception and action algorithm for autonomous power line inspection that does not require prior information about the power line infrastructure, such as the location of the power lines and masts. Our method plans and tracks a trajectory that maximizes the visibility of the power line in the onboard camera view and, at the same time, can safely avoid obstacles such as the power masts. We achieve this by developing a perception-aware Model

Predictive Controller (MPC) that includes two perception objectives: line tracking and collision avoidance. To detect the power lines, we propose a novel perception module that extends the deep-learning–based object detector in [6] to the case of power line detection. The perception module is trained only on synthetic data and transfers zero shots to real-world images of power lines without fine-tuning. In this way, we overcome the problem of the limited amount of annotated data for supervised learning.

8.2.1.1 System design and validation

Our system is based on the MPC formulation for quadrotors proposed in [7]. We extend this MPC by including two new perception objectives: one for line tracking and another one for collision avoidance.

Line Tracking Objective: The purpose of this objective is: (1) to keep the power line in the center of the image, to maximize data quality for visual inspection, and (2) to keep a safe distance from the power lines. Specifically, we convert the positions of line endpoints from a world frame into polar coordinates in the image frame. We add a perception cost to force the lines to be centred in the image frame. We also introduce an objective to maintain the desired distance between the drone and the power lines.

Obstacle Avoidance Objective: For obstacle avoidance, we employ a collision cost and a collision constraint [8]. The collision cost is determined using a logistic function that considers the distance between the drone and the detected obstacles. The collision constraint is established as a probabilistic chance constraint to account for uncertainty in drone and obstacle states. The goal is to ensure that the collision probability with an obstacle remains below a specified threshold. Obstacles are modeled as ellipsoids. The uncertain nature of the positions is accounted for by assuming Gaussian distributions for the position of both the quadrotor and the obstacle.

8.2.1.2 Line detection and tracking

We propose a deep-learning-based power line detector based on the object detector [6]. The detector takes a single RGB image as input and outputs end points of the detected power lines in pixel coordinates. The centre patch of each detection is matched with the prediction of the previous patch using the Hungarian method [9]. We use a KLT tracker [10] to perform tracking. The final output is the tracked lines endpoints which are given to the MPC. To overcome the problem of the lack of datasets containing labelled images of power lines, we created a new synthetic dataset for power line detection based on the Flightmare simulator [11].

Figure 8.1. A quadrotor performing power line inspection in a power line test environment with three masts (labelled: A, B, and C) using our proposed approach. The original rough reference trajectory is depicted in yellow and the drone deviates from it while avoiding obstacles (the power masts) and keeping the power line visible in the field of view of the onboard camera.

Our model is trained on circa 30k simulated images. We show real-world deployment without any fine-tuning on real images.

8.2.1.3 Validation

We validated our system on a custom-made quadrotor [12]. Our quadrotor is equipped with an Intel RealSense T265 tracking camera and an Intel Realsense D435i depth camera. The onboard computer is a Nvidia Jetson TX2. We use the VIO algorithm from the tracking camera to obtain an estimate of the 6-DoF pose of the quadrotor and the depth camera to obtain RGB images for the line detection algorithm and depth measurements for the collision avoidance algorithm. All the components of our system run on the onboard computer in real time. A top view of one of our experiments is represented on Figure 8.1.

8.2.2 Long-Range Powerline Mapping

The need for cost-effective solutions for powerline mapping and inspection has motivated the development of aerial robots based on unmanned autonomous helicopters, Vertical Take-Off and Landing (VTOL), or multi-rotors. Despite the shorter flight endurance of multi-rotors, in the last years, there is significant research to provide multi copters with the capacity to perch on powerlines for recharging batteries, see e.g., [13]. These robots are equipped typically with visual cameras and LIDARs. Although they implement different autonomous functionalities, in all cases, powerline mapping is performed in a non-autonomous way that is structured in two stages. First, the robots fly through teleoperation or use predetermined missions to register measurements without performing any online processing. Second, in the office/lab, inspection is performed by offline, analysing the collected data using processing and artificial intelligence methods. In this procedure, the operators cannot be sure that the obtained maps have sufficient resolution or accuracy, potentially requiring repeating the data collection flights weeks after the data collection flights, hence involving delays, costs and operational problems.

This mapping system and methods developed in H2020 AERIAL-CORE enable the fully autonomous and online building of accurate 3D maps suitable for long-range powerline inspection. The developed system builds online and fully on board the robot an accurate (mean error of <5 cm) 3D map of the area surrounding the powerline, enabling measuring distances between vegetation and the electrical system without requiring offline processing. The developed system is based on a fully autonomous multi-rotor aerial robot capable of performing Beyond Visual Line of Sight (BVLS) flights. The robot flies 10 m above the powerline and towers. It follows the powerline describing a trajectory specified as a set of waypoints defined by the operator based on existing maps of the electric system. It includes a GNSS-based Trajectory tracker module that gives the commands to the robot autopilot. During navigation, the point clouds from the 3D LiDAR are collected and logged and processed to online build an accurate 3D map of the environment. The robotic system is also endowed with methods that process the obtained LiDAR-based geometrical map to segment the map points into four classes *Powerline*, *Tower*, *Vegetation*, and *Soil*.

The mapping engine is based on FAST-LIO2 but enhanced to integrate GNSS measurements in the Update of its iterated Kalman Filter scheme to improve robustness in scenarios with poor geometrical information content. FAST-LIO2 works directly with full LiDAR scans instead of with feature points as the other methods; it is based on error-state manifold-based iterative Kalman Filter, which enables reducing the memory footprint (suitable for long-range missions) and has moderate CPU consumption that enables online execution without discarding data. Refer to [15] for further details. Autonomous real-time 3D segmentation was performed by first using the reflectivity component of LiDAR points to differentiate whether the object is metallic (*Powerline* or *Tower*) or not (*Vegetation* or *Soil*). Second, using Principal Component Analysis (PCA), the geometrical information of the point clouds of objects classified as metallic or non-metallic was used to differentiate between *Powerline* or *Tower* (for objects classified as metallic) and *Vegetation* or *Soil* (for objects classified as non-metallic). Refer to [16] for further details.

The method was implemented on the *LR-M* aerial robot developed by the GRVC Robotics Lab at the University of Seville, see Figure 8.2-left. *LR-M* is based on the DJI Matrice 600 hexacopter platform, and its main sensors are a Livox Horizon 3D solid-state LiDAR pointing at a pitch angle of -40 degrees and an

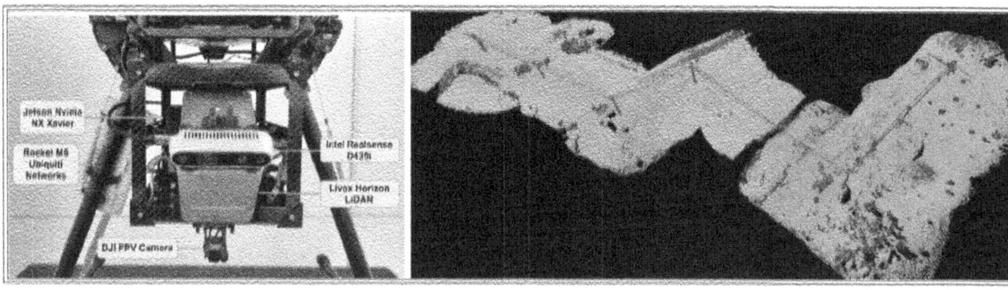

Figure 8.2. (Left) Sensors and computers on board LR-M robot used in the experiments. (Right) Powerline map built by LR-M in Alcala scenario (Sevilla) covering >1.8 km.

Intel RealSense RGB camera mounted with the same orientation for user visualization. In addition, it is equipped with a Jetson NVIDIA NX Xavier for online computation and logging. The Livox Horizon is a high-performance solid-state 3D LiDAR with a non-repetitive horizontal scanning pattern. It has a field of view of 81.7° × 25.1°, a detection range of 260 m, an angular precision of 0.05°, and a point rate of 240,000 pts/s. The mapping system implemented on *LR-M* was validated in experiments performed at (i) School of Engineering of Seville (ETSI, Seville, Spain), (ii) Burguillos (Seville, Spain), and (iii) Alcala de Guadaira (Seville, Spain). Figure 8.2-right shows the powerline map resulting in long-range mapping mission experiments performed in the Alcala scenario.

8.3 Aerial Robotic Manipulation for Maintenance

In the development of aerial manipulation robots intended to conduct maintenance operations involving physical interaction with the environment using tools or devices, like in the installation of bird flight diverters on power lines considered in AERIAL-CORE, it is possible to identify three design approaches, as described in [4], depending on the main constraint to be satisfied: weight or force required to manipulate the devices or tools, maximum payload capacity provided by the aerial platform, or level of dexterity needed to conduct the task.

The comparative case study presented in [17] identifies three modes of operation for the aerial robot depending on the way the manipulation task is conducted: (1) while flying as occurs with the linear actuator platform developed in [18] and as typically considered in most aerial manipulation works [4], (2) in grabbing conditions using one arm while the other operates, exploiting the passive accommodation capabilities of the compliant arm [19], and (3) perching the platform or deploying the manipulator on the workspace, detaching it from the aerial platform, used for its transportation and retrieval [20].

The dichotomy between developing general-purpose aerial manipulators or platforms to conduct specific tasks more efficiently or more suitably has been addressed

Figure 8.3. Linear Actuator Platform (LAP, up-left), Dual Arm Platform with Cart (DAP-C, down-left), and Main Local Manipulation Platform (MLMP, right) developed as part of the AERIAL-CORE project.

within the AERIAL-CORE project, resulting in the three different prototypes shown in Figure 8.3. These are Linear Actuator Platform (LAP, up-left) [18], Dual Arm Platform with Cart (DAP-C, down-left) [20], and the Main Local Manipulation Platform (MLMP) which is a high payload capacity multi-rotor, with a maximum take-off weight of 45 kg, providing electrical shielding against high voltage power lines, equipped with a high payload capacity (5 kg) robotic arm capable of conducting the installation of different types of devices considered within the project.

Subsection 8.3.1 describes a high payload manipulator for large platforms, particularly the MLMP. Subsection 8.3.2 summarizes the main characteristics of the Linear Actuator Platform and the Dual Arm Platform with Cart. Finally, Subsection 8.3.3 describes the *MLMP*.

8.3.1 High Payload Light Arm for Aerial Manipulation Tasks

The tasks to be conducted by the manipulator entail installing and removing different types of devices from the power lines. These include bird diverters (helical and clip diverters) used to prevent bird collisions, threatening birds and placing electrical cable spacers. Additionally, the power lines are exploited to improve the efficiency of the aerial system: recharging stations may need to be installed to extend the UAVs' battery life.

The proposed manipulator is an anthropomorphic arm featuring six degrees of freedom (DoFs), accompanied by a gear-based spherical wrist that supports the end effector. This configuration is essential to achieve a superior level of dexterity,

Figure 8.4. Real world experiment tests case: installation of a recharging station on a power line. (a) Approaching, (b) tool inserting, (c) clamping, (d) releasing. A video of this process can be seen at http://y2u.be/kvChEYhnxtY.

enabling the arm to accomplish all the designated tasks.The last joint of the arm has been thoughtfully designed with a versatile flange, allowing for easy attachment of different end-effectors specifically tailored to carry out specific tasks.

From the first prototype, the arm has been designed to weigh 3 kg, making it lightweight and agile. Despite its compact size, it boasts a payload capacity of 5 kg. This remarkable payload/weight ratio of 1.67 sets it apart from other similar solutions proposed in the literature (e.g., 0.5 for Haddington Dynamics Dexter HDI, which has a similar mechanical structure). The first version of the spherical wrist was built in plastic; even if the mechanism worked correctly, the material strength and rigidity were insufficient to perform all the required tasks. Also, using belt transmissions in the arm introduced inherent challenges related to elasticity. Displacements of the end effector caused by elasticity made it incapable of autonomously executing assigned motion tasks.

The latest version of the arm has solved previous issues. The wrist has been remade in steel (AlSi10Mg DSLM), achieving better stability and performance with a negligible increase in weight. The low-level control is implemented on an onboard microcontroller, making the system stand-alone. It can be teleoperated and used autonomously, providing an additional communication channel to an external PC for high-level controls. By leveraging the data provided by two IMUs, it is possible to calculate the precise position of the end effector, applying corrective measures and allowing the arm to execute tasks even in the presence of elasticity successfully. The proposed system has proven to complete all the assigned tasks in a mock-up scenario and on a real powerline, as illustrated in Figure 8.3-right and in Figure 8.4.

8.3.2 Linear Actuator Platform and Dual Arm Platform

The Linear Actuator Platform (LAP) [18] is developed for the installation of a particular model of clip-type bird flight diverters, extensively used on the Spanish power grid, shown in Figure 8.5, that replicates the shape of birds of prey to fear

Figure 8.5. Linear Actuator Platform (left and middle) used for the installation of clip-type bird flight diverters (up-left), and Dual Arm Platform with Cart (right) for its deployment on the power line and installation of customized clip-type bird diverters.

birds approaching the power lines. The operation requires the application of very high forces (up to 100 kg) using a Firgelli linear actuator equipped with a clamp mechanism that isolates the aerial platform from the exerted pushing force by creating a closed kinematic chain when acting on the cable. The installation mechanism integrates the onboard control electronics, battery and a radio link, resulting in a compact device weighing 2 kg capable of conducting the installation of the device in 12 seconds. The motion constraint imposed by the clamp mechanism during the operation requires a certain level of accommodation of the aerial platform to unavoidable small position deviations, taking into account the typical positioning accuracies that can be achieved outdoors, incorporating for this purpose a passive spherical joint to attach the mechanism at the base of the multi-rotor to avoid destabilizing it due to the interaction wrenches exerted on flight.

The Dual Arm Platform [19] and Dual Arm Platform with Cart (DAP-C) [20] consist of a human-like and human-size dual-arm manipulator providing dexterous manipulation capabilities, with four joints for end effector positioning that allow replication of the bimanual dexterity of human workers. The very low weight of the arms (3 kg) facilitates its aerial transportation and deployment on the power line using medium-scale aerial platforms like the one shown in Figure 8.5(right). At the same time, the rolling base provides an energy-efficient way of moving along linear infrastructures. The arms integrate a compact spring-lever transmission in all joints to provide mechanical compliance, which contributes to protecting the servo actuators from impacts and overloads and results in safer physical interactions with the environment. In general, the idea of deploying the manipulator on the power line and detaching the multi-rotor from it results more convenient than operating on flight in terms of positioning accuracy, reliability, and energy efficiency. However, the realization of the manipulation operation is constrained by the kinematic configuration of the dual arm system concerning the power line, whereas the realization of the aerial manipulation on flight extends the effective workspace of the arms.

The use of dual manipulators presents two significant benefits, explored in AERIAL-CORE. On the one hand, taking into account the positioning accuracy required to conduct the manipulation task (in the range of 1 cm), it is interesting to consider the application of one of the arms to measure the position of the aerial robot concerning the workspace when the arm is grabbed to a fixed point, taking benefit of the passive compliance of the arm, as reported in [19]. On the other hand, the symmetry of a dual-arm manipulator makes it possible to maintain the equilibrium on linear infrastructures like cables or pipes as long as the centre of mass of the manipulator is below the perching point.

8.3.3 The Main Local Manipulation Platform

The Main Local Manipulation Platform (MLMP), depicted in Figure 8.6, represents the fourth demonstrator of AERIAL-CORE's aerial robotic manipulation module. Designed as a versatile system, it can install, remove, or manipulate various equipment typically employed in electrical infrastructure. During the project, four devices were targeted for this demonstrator: clip-type bird diverters, cable separators, and two types of drone charging stations [21]. The operating principle of the platform consists of navigating to an energized power line, perching on it, and turning off the motors to conserve energy. Afterwards, it moves along the line while installing the equipment.

Considering that the MLMP has been designed as a general-purpose platform, it should be capable of installing various types of devices without requiring extensive modifications between operations. To achieve this goal, it was decided to perform

Figure 8.6. MLMP using the robotic arm to charge its batteries from a drone charging station on a controlled environment.

the manipulation with a robotic arm with specific end effectors for each device rather than a general one. In this way, the versatility of the arm is balanced with the efficiency and customization of the end effectors, significantly reducing the difficulty of the manipulation. As mentioned above, the robotic arm was designed with six degrees of freedom to ensure good manoeuvrability, with the main motors located in the base and connected to the joints by a belt system. This ensures that the drone's centre of mass is not excessively disturbed, which could cause problems during flight. On the software side, both joint and cartesian controllers were developed, using a standard gamepad to command the references.

Another challenge was the perching process from below, integrating a specially designed mechanism with a dedicated perception and control system. The mechanism allows the drone to perch on the cable smoothly and robustly using only one servo motor connected to a proprietary hook design. It also includes two pulleys attached to DC motors, allowing the drone to move along the cable (see Figure 8.6). Additionally, to minimize the difficulty of inserting the cable inside the mechanism, a V-shaped structure capable of folding itself using two linear motors was integrated on top of the mechanism, acting as a guide for the cable when the perching manoeuvre is being performed, and pressing the cable when it is inside the mechanism to increase the stability of the drone in the line.

The perching software is built around two sensors: a pair of two-dimensional LIDARs and an RTK GPS. The drone initiates the mission with prior knowledge of the approximate GPS position and elevation of the line. It flies with a standard navigation system complemented by the centimetric precision of the RTK GPS. Once there, the drone initiates the perching phase by approaching under the cable while the perception system, which employs both LIDARs, attempts to detect it. When the line is detected, the drone begins to ascend. If the cable is detected inside the mechanism, the rotation is activated, and the cable is locked, completing the perching phase. Finally, the drone turns off its motors and starts the manipulation phase.

With the manipulation and perching challenges solved, the final step was to design a platform that was robust enough to integrate the robotic arm and all the necessary sensors and capable of operating in contact with a live power line with hundreds of kilovolts of tension. To address the two primary issues that arise when a drone touches a high-voltage power line, which are the voltage itself and the electromagnetic interference (EMI), the solution adopted was to build the drone with a closed metal frame and place all the electronics inside with their grounding connected to it. To test this solution, a prototype was made and tested by touching a 125 kW power line with excellent results. In all attempts, the drone touched the line without any problem, and neither the sensors nor the control suffered any interference or disturbances due to the line.

The final platform was developed by combining all the above technologies, and numerous integration and validation experiments were performed in controlled and realistic environments. For the controlled experiments, a mock-up cable was installed at the ATLAS Flight Test Centre in Villacarrillo (Spain). All the devices were installed, and several perching attempts were made with very positive results. The realistic experiments were also conducted at ATLAS, but in a real power line located close to the facilities. A complete operation of installing a bird diverter was performed, including autonomous navigation to the line, autonomous perching, teleoperated installation and removal of the device, releasing and returning to home.

In conclusion, the MLMP represents a breakthrough in power line inspection and maintenance operations. It is the world's first drone capable of perching on live high-voltage power lines and installing multiple types of devices using a six-degrees-of-freedom robotic arm while moving along the cable.

8.3.4 Magnetic Gripper with Charging Function for Drones on Power Lines

One limitation of drones in inspecting transmission lines is the short flight time. A promising solution is to charge the battery from the magnetic field around the power lines [22–28], allowing drones to operate continuously without returning to the base station. The charging time varies with the power line current, ranging from less than an hour to a couple of hours. In addition to the charging circuit, the grasping systems are equipped for the drone to perch on the cable. Therefore, to increase the flight time, the weight of these systems needs to be optimized to reduce the load on the drone.

Usually, the grasping systems include three components: current transformer (CT), gripper, and actuators (Figure 8.7). The current transformer, made of a magnetic core and coil, harvests magnetic energy. The gripper is employed to hold

Figure 8.7. Typical grasping system of a drone charging from the power line [23].

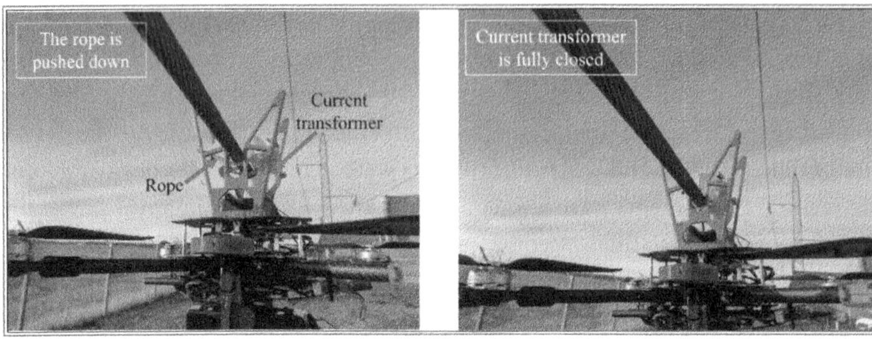

Figure 8.8. An optimized grasping system with only one current transformer [27].

the drone on the line. The actuators are typically motors that open and close the magnetic core and the gripper. Besides the function of harvesting energy, CT can act as an electromagnet when it is magnetized by the AC magnetic field. Therefore, it is possible to take advantage of this magnetic force to hold the drone on the cable and remove the gripper and actuators to minimize the payload on the drone.

The magnetic gripper in Figure 8.8 enables the drone to perch on the cable with only one CT [27]. When the drone flies upwards, the power cable pushes down on the rope that links the two halves of the magnetic core, closing the gripper. The CT maintains the grip thanks to the magnetic force and harvests energy simultaneously. The core is open by removing the magnetic force, and the drone falls off by itself. However, it is essential to know how much force is needed to hold the drone, as the magnetic force depends on the level of power line current. Assuming the fringing flux is negligible, the magnetic force at one contact surface of the core is calculated as below.

$$F = \frac{B^2 A_{CORE}}{2\mu_0} \tag{8.1}$$

Where B is the core's flux density, A_{CORE} is the cross-sectional area of the core (m^2), and μ_0 is the permeability of the air. As A_{CORE} and μ_0 are fixed values, the force varies with B, which can be electrically controlled by adjusting the magnetizing current I_m, as shown in the equation below.

$$B = \frac{N I_m \mu_0 \mu_e}{l_m} \tag{8.2}$$

Where N is the number of winding turns, μ_e is the core's effective permeability, and l_m is the magnetic path length of the core (m). The magnetizing current I_m is the amount of current used to magnetize the core, as shown in the equivalent circuit of CT in Figure 8.9-left, where I_p is the primary current (or power line current), I_s is the secondary current, and I_{Load} is the load current. The directions of

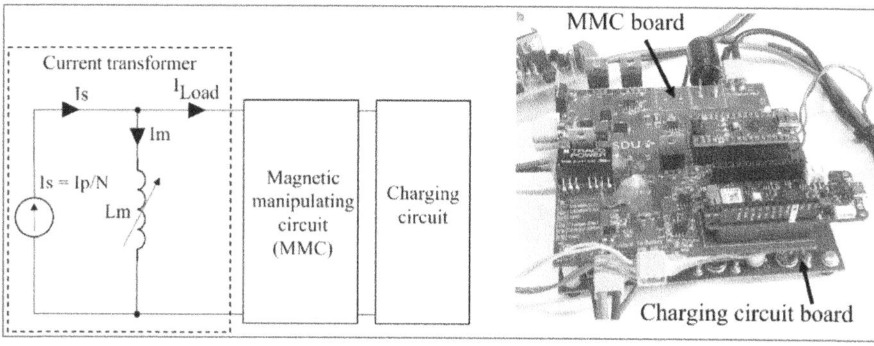

Figure 8.9. Left: The equivalent circuit of CT in connection with MMC and the charging circuit, Right: MMC and the charging circuit in practice.

currents change over time due to the AC current on the line. L_m is the non-linear magnetizing inductor.

Since I_m depends on I_p, I_m can be controlled based on the level of I_p. This can be done by a magnetic manipulating circuit (MMC), as shown in Figure 8.9-right. When there is no power line current or low, DC power from the battery is utilized to maintain the holding force. In this case, I_m consists of AC current from I_p and DC current from I_{Load}. When I_p is high enough, MMC starts charging the battery while keeping enough I_m. When the charging process is completed and I_p is still high, the magnetic field from the power cable is exploited to magnetize the core instead of using DC power from the battery to save energy. MMC controls I_p to increase I_m instead of I_{Load} as in the charging phase.

8.4 Human-Machine Interfaces

Some operations during inspection and maintenance are difficult to perform autonomously, and sometimes it is necessary for the operator to take control of the platform to perform such operations.

There are several existing approaches to the teleoperation of drones and manipulators. As robots intended for these applications become more complex and powerful, additional efforts are needed to implement interfaces that are both effective and convenient for most users [29]. However, standard interfaces such as remote controllers still fail to achieve this goal and require significant time and effort to be mastered by inexperienced users [30, 31]. For this reason, several new remote-control methods for drones and aerial manipulators have been developed to provide operators with more intuitive operation and feedback to all their senses. These include control methods based on the operator body pose, telemanipulation of robotic arms based on a replica of the controlled system, and control based on operator voice

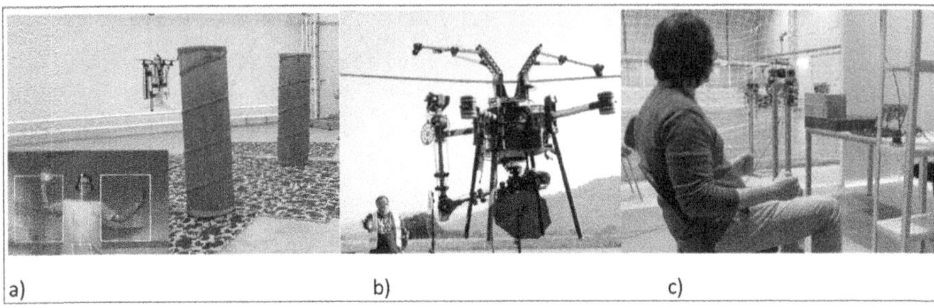

Figure 8.10. Different human-machine interfaces: (a) operator controls UAV based on human pose estimated from camera image; (b) operator controls robotic arm on MLMP based on human pose estimated from IMU measurement; (c) operator controls light dual arm based on human pose estimated from camera image.

commands. Feedback to the operator is provided through various interfaces that provide visual, haptic and acoustic information.

For control based on operators' body pose, we developed an IMU-based human pose estimation and a camera-based human pose estimation, which was then mapped to the different control laws. In [32] we presented a UAV control strategy which relies on estimating the operator's body posture from the camera image, for which a user study was conducted and assessed based on the NASA Task Load Index. The method was implemented and tested on the morphing rotary /fixed wing platform (see Figure 8.10.a)). The concept of simultaneous control of UAV and manipulator based on the operator's body pose from the camera image was presented in [33], and a similar idea was tested based on IMU estimated body pose. A similar control method with an operator equipped with IMUs was also tested in a realistic scenario with the MLMP to control the robotic arm, as shown in Figure 8.10.b.

Three different teleoperation interfaces were investigated for the teleoperation of anthropomorphic dual arms with very low weight. A comparative performance evaluation was performed in [34]. The three methods include a visual human pose estimation system that maps the pose of the human arm to the pose of the robotic arm (see Figure 8.10.c), a leader-follower scheme that uses a scaled-down dual arm that can directly transfer the joint positions of the leader arm to the follower arm, and a 6-DOF (degrees of freedom) joystick.

To provide individualized control for each operator, a learning-based approach was developed and used with different interfaces that include direct mapping of operator-attached IMU measurements, motion capture data, and hand posture-based control with hand and finger motion controllers. In this approach, the operator first attempts to mimic previously defined trajectories, and his or her movements are used to teach the system control commands. After the learning phase, the system maps the control of different movements to specific control actions, and the operator can control the system using his own control commands.

Various interfaces have been used to provide feedback to the operator. The primary method of feedback is visual, for which computer screens, head-mounted displays, and smart glasses have been used. Unlike computer screens, head-mounted displays and smart glasses provide an immersive experience for the user. Different graphical interfaces were used as needed, but the primary channel of information transmission is the first-person perspective from the teleoperated object. In addition to visual feedback, auditory and haptic feedback were also used. Different combinations of control and feedback methods were used to conduct experiments and user studies on the usability and intuitiveness of the developed methods.

8.5 Formation Control of Safety Aerial Co-Workers

Work at height, such as on power line maintenance, is physically and mentally demanding. Human operators have limited mobility for handling the required tools and situational awareness around the workplace. To this end, the Aerial Co-Workers (ACW), an Unmanned Aerial Vehicle (UAV) helping the worker, can significantly improve the effectiveness of the work and its safety. Three types of Aerial Co-workers can be employed in these situations. Physical ACW [35] that can physically interact with human workers, e.g., to deliver tools or components. Inspection ACW [36] can provide support to human operators in acquiring views of the power tower that are not easily accessible. Finally, the Safety ACWs, which we focus on in this section, can be tasked to monitor the human workers performing maintenance on the critical infrastructure of the powerlines. By doing so, the Safety ACWs reduce the probability of severe accidents leading to workers' injury while enabling better situational awareness around the worker, similar to the Inspection ACW.

A natural way to acquire more information about the worker's surroundings and thus his safety is to employ multiple robots that form a team of Safety ACWs. The team is thus tasked to keep a formation around the worker, monitor the worker with onboard sensors, and change the appearance when commanded by the worker. This requires from the Safety-ACWs formation to perform planning and control strategies to keep the robots within the formation at a given mutual distance from each other and from the human worker. This all while observing the worker with onboard sensors and avoiding collisions with the environment. Moreover, the formation of Safety-ACWs is controlled using gesture recognition [37] to change both lateral and vertical relative viewpoints of the worker.

The Safety-ACWs uses leader-follower formation scheme, similar to our work [38], where one ACW UAV is the formation leader that carries the onboard

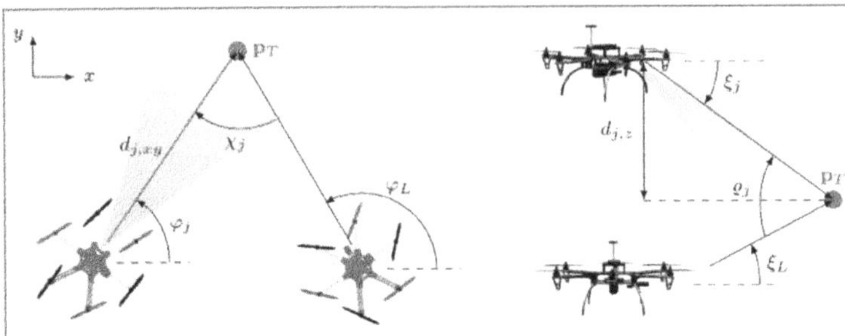

Figure 8.11. The leader-follower scheme of Safety Aerial Co-Worker formation.

sensors to detect the human worker and its gestures. The follower ACW then keeps a predefined formation with the leader relative to the detected position of the worker (shared within the team), and also keeps heading towards the worker. The reference trajectories for the follower UAVs are computed to achieve a desired leader-follower formation around the target. The desired position of the followers is influenced by the corresponding leader position $\mathbf{p}_L$ and camera orientation $\mathbf{o}_L$, the target position $\mathbf{p}_T$, the desired follower angles of j-th follower χ_j and ϱ_j, and the desired distance of the follower to the worker d_j.The desired position of j-th follower $\mathbf{p}_j$ is then given by the equation:

$$\mathbf{p}_j = \mathbf{p}_T + d_j \begin{bmatrix} -\cos(\varphi_j)\cos(\xi_j) \\ -\sin(\varphi_j)\cos(\xi_j) \\ \sin(\xi_j) \end{bmatrix}, \quad (8.3)$$

where $\varphi_j = \varphi_L + \chi_j$ and $\xi_j = \xi_L + \varrho_j$ are desired follower angles relative to the camera (see Figure 8.11).

Both the leader ACW and the followers use trajectory planning to reach a desired positions with respect to the worker as described above. The trajectory planning uses an initial path-planning phase where a collision-free path is planned using Jump Point Search (JPS) algorithm introduced for 3D in [39]. The path is then used to find a safety flight corridor between the current UAV position and the desired position. The flight corridor is a convex decomposition of the free space found on a map of the environment as described in [39]. Finally, given a collision-free path P and its corresponding safe corridor S, a final optimal trajectory is computed using a Quadratic Programming (QP) formulation of trajectory planning in a receding horizon. The particular optimization problem minimizes both the control effort and the error from the desired path. This QP constrains both the UAV inputs and speed, while it also limits the UAV positions to be within the convex polyhedron representing its flight corridor. This ensures the final trajectories of the team

Figure 8.12. Real demonstration of team of three Safety-ACWs controlled by gestures of the human worker.

are collision-free while keeping the Safety ACW-formation. A real demonstration of the system is shown in Figure 8.12.

8.6 Conclusions and Future Work

The AERIAL-CORE project has achieved relevant progress on intelligent robotic technologies and systems applied to electrical power line inspection and maintenance.

The Chapter has presented intelligent systems for long-range inspection of the power lines, by including the perception-aware model predictive control for the autonomous tracking of the line and the long-range on-board autonomous mapping of the power line and the vegetation around the line.

It also presented the application of aerial robotic manipulation technologies by including several aerial manipulation platforms and a high payload arm. These technologies are being used to install bird-diverters and also a battery charger station allowing UAVs to operate continuously without returning to the base station. The main characteristics of the magnetic gripper were also presented.

The Chapter also included human-machine interfaces and teleoperation technologies to operate the aerial platform and the arms.

Finally, the Chapter presents a team of aerial robot with formation control to monitor the safety of a human operator.

The Chapter included prototypes tested in real electrical lines. Preliminary integration and validation experiments have been performed in May 2022 and May 2023 in the ATLAS Airfield with medium and high voltage electrical lines.

The experiments have pointed out the industrial interest of the presented systems for inspecting and maintaining the power grid.

Other AERIAL-CORE technologies that will be demonstrated in the next months in ATLAS include morphing of the aerial platforms, multi-aerial robot planning techniques, learning and gesture recognition in the interaction with humans and the integration of the AERIAL-CORE system.

It is expected that several AERIAL-CORE technologies will be industrialised and applied at short term for the inspection of the electrical grid.

Acknowledgements

This work was supported by the AERIAL-CORE Project of the European Commission under Grant H2020-2019-871479.

References

[1] EU Commission, "Drone strategy 2.0: Creating a large-scale European drone market", https://ec.europa.eu/commission/presscorner/detail/en/ip_22_7076, accessed: 2023-09-07.

[2] CNN, "California's second-largest wildfire was sparked when power lines came in contact with a tree", https://edition.cnn.com/2022/01/05/us/dixie-fire-power-lines-cause-pge, accessed: 2023-09-07.

[3] F. Ruggiero, V. Lippiello, A. Ollero. "Aerial manipulation: A literature review" in IEEE Robotics and Automation Letters 3(3), 1957–1964.

[4] A. Ollero, M. Tognon, A. Suarez, D. Lee and A. Franchi, "Past, Present, and Future of Aerial Robotic Manipulators", in IEEE Transactions on Robotics, vol. 38, no. 1, pp. 626–645, Feb. 2022, doi: 10.1109/TRO.2021.3084395.

[5] J. Xing, G. Cioffi, J. Hidalgo-Carrio, D. Scaramuzza, "Autonomous Power Line Inspection with Drones via Perception-Aware MPC", IEEE/RSJ Int. Conf. Intel. Rob. and Syst. (IROS), 2023, Detroit, USA.

[6] J. Redmon, S. Divvala, R. Girshick, and A. Farhadi, "You only look once: Unified, real-time object detection", IEEE Conf. on Computer Vision and Pattern Recognition (CVPR), 2016, Las Vegas, USA.

[7] D. Falanga, P. Foehn, P. Lu, and D. Scaramuzza, "Pampc: Perception-aware model predictive control for quadrotors", IEEE/RSJ Int. Conf. Intell. Robot. Syst. (IROS), 2018, Madrid, Spain.

[8] H. Zhu and J. Alonso-Mora, "Chance-constrained collision avoidance for mavs in dynamic environments", IEEE Robot. and Autom. Let. (RA-L), vol. 4, no. 2, pp. 776–783, 2019.

[9] H. W. Kuhn, "The hungarian method for the assignment problem" Naval research logistics quarterly, vol. 2, no. 1–2, pp. 83–97, 1955.

[10] B. D. Lucas and T. Kanade, "An iterative image registration technique with an application to stereo vision", in 7th Int. Joint Conf. on Artificial Intel. (IJCAI), vol. 2, 1981, pp. 674–679.

[11] Y. Song, S. Naji, E. Kaufmann, A. Loquercio, and D. Scaramuzza, "Flightmare: A flexible quadrotor simulator", in Conf. on Robot Lear. (CORL), 2021, Virtual.

[12] P. Foehn, E. Kaufmann, A. Romero, R. Penicka, S. Sun, L. Bauersfeld, T. Laengle, G. Cioffi, Y. Song, A. Loquercio, and D. Scaramuzza, "Agilicious: Open-source and open-hardware agile quadrotor for vision-based flight", Science Robotics, vol. 7, no. 67, 2022.

[13] J. L. Paneque, J. R. Martínez-de Dios, A. Ollero, D. Hanover, S. Sun, A. Romero, and D. Scaramuzza, "Perception-aware perching on powerlines with multirotors", IEEE Robotics and Automation Letters 7(2), 3077–3084, 2022.

[14] W. Xu, Y. Cai, D. He, J. Lin, and F. Zhang, "Fast-lio2: Fast direct lidar-inertial odometry", IEEE Transactions on Robotics, pp. 1–21, 2022.

[15] J. L. Paneque, V. Valseca, J. R. Martínez-de Dios, and A. Ollero, "Autonomous Reactive LiDAR-based Mapping for Powerline Inspection", Int. Conf. on Unmanned Aircraft Systems (ICUAS), 962–971, 2022.

[16] V. Valseca, J. Paneque, J. R. Martínez-de Dios, and A. Ollero, "Real-time LiDAR-based Semantic Classification for Powerline Inspection", Int. Conference on Unmanned Aircraft Systems (ICUAS), 478–486.

[17] A. Suarez and A. Ollero, "Dual Arm Aerial Manipulation While Flying, Holding and Perching: Comparative Case Study", in 2022 Iberian Robotics conference. Springer Cham International Publishing, pp. 259–270, 2022.

[18] A. Suarez, H. Romero, R. Salmoral, J. A. Acosta, J. Zambrano and A. Ollero, "Experimental Evaluation of Aerial Manipulation Robot for the Installation of Clip Type Bird Diverters: Outdoor Flight Tests", *2021 Aerial Robotic Systems Physically Interacting with the Environment (AIRPHARO)*, Biograd na Moru, Croatia, 2021, pp. 1–7, doi: 10.1109/AIRPHARO52252.2021.9571029.

[19] A. Suarez, P. J. Sanchez-Cuevas, G. Heredia, and A. Ollero, "Aerial Physical Interaction in Grabbing Conditions with Lightweight and Compliant Dual Arms", Applied Sciences, vol. 10, no. 24, p. 8927, Dec. 2020, doi: 10.3390/app10248927.

[20] A. Suarez, S. Rafee-Nekoo, and Anibal Ollero, "Ultra-lightweight anthropomorphic dual-arm rolling robot for dexterous manipulation tasks on linear infrastructures: A self-stabilizing system", Mechatronics, vol. 94, 2023.

[21] A. M. Muñoz-Gómez, J. M. Marredo-Píriz, J. Ballestín-Fuertes, and J. F. Sanz-Osorio, "A Novel Charging Station on Overhead Power Lines for Autonomous

Unmanned Drones", Appl. Sci. 2023, 13, 10175. doi: 10.3390/app13181 0175.

[22] R. Kitchen, N. Bierwolf, S. Harbertson, B. Platt, D. Owen, K. Griess-mann, and M. A. Minor, "Design and evaluation of a perching hexacopter drone for energy harvesting from power lines", in 2020 IEEE/RSJ International Conference on Intelligent Robots and Systems (IROS). IEEE, 2020, pp. 1192–1198.

[23] N. Iversen, O. B. Schofield, L. Cousin, N. Ayoub, G. Vom Bögel, and E. Ebeid, "Design, integration and implementation of an intelligent and self-recharging drone system for autonomous power line inspection", in 2021 IEEE/RSJ International Conference on Intelligent Robots and Systems (IROS). IEEE, 2021, pp. 4168–4175.

[24] N. Iversen, A. Kramberger, O. B. Schofield, and E. Ebeid, "Novel power line grasping mechanism with integrated energy harvester for uav applications", in 2021 IEEE International Symposium on Safety, Security, and Rescue Robotics (SSRR). IEEE, 2021, pp. 34–39.

[25] W. Stewart, D. Floreano, and E. Ebeid, "A lightweight device for energy harvesting from power lines with a fixed-wing uav", in 2022 International Conference on Unmanned Aircraft Systems (ICUAS). IEEE, 2022, pp. 86–93.

[26] G. Vom Bögel, L. Cousin, N. Iversen, E. S. M. Ebeid, and A. Hennig, "Drones for inspection of overhead power lines with recharge function", in 2020 23rd Euromicro Conference on Digital System Design (DSD). IEEE, 2020, pp. 497–502.

[27] V. Duong Hoang, A. Kramberger, and E. Ebeid, "Adaptive and fail-safe magnetic gripper with charging function for drones", in 2023 IEEE/RSJ International Conference on Intelligent Robots and Systems (IROS). IEEE, 2023 (accepted).

[28] V. Duong and E. Ebeid, "Advanced magnetic energy harvester for charging drones from overhead powerlines", in The 25th European Conference on Power Electronics and Applications (EPE '23 ECCE Europe). IEEE, 2023 (accepted).

[29] J. Casper and R. R. Murphy, "Human-robot interactions during the robot-assisted urban search and rescue response at the World Trade Center", IEEE Transactions on Systems, Man, and Cybernetics, Part B: Cybernetics, vol. 33, no. 3, pp. 367–385, 2003.

[30] J. M. Peschel and R. R. Murphy, "On the human-machine interaction of unmanned aerial system mission specialists", IEEE Transactions on Human-Machine Systems, vol. 43, no. 1, pp. 53–62, 2013.

[31] J. Y. Chen, E. C. Haas, and M. J. Barnes, "Human performance issues and user interface design for teleoperated robots", IEEE Transactions on Systems, Man and Cybernetics Part C: Applications and Reviews, vol. 37, no. 6, pp. 1231–1245, 2007.

[32] Zorić, F., Vasiljević, G., Orsag, M., and Kovačić, Z., "Towards Intuitive HMI for UAV Control", In 2022 International Conference on Smart Systems and Technologies (SST), pages 325–332, 2022.

[33] Zorić, F., and Orsag, M. "H2AMI: Intuitive Human to Aerial Manipulator Interface", In 2023 International Conference on Unmanned Aircraft Systems (ICUAS), pages 1226–1232, 2023.

[34] Zorić, F., Suarez, A., Vasiljević, G., Orsag, M., Kovačić, Z., Ollero, "A. Performance Comparison of Teleoperation Interfaces for Ultra-Lightweight Anthropomorphic Arms". Accepted for 2023 IEEE/RSJ International Conference on Intelligent Robots and Systems (IROS 2023), 2023.

[35] A. Afifi, M. van Holland, and A. Franchi, "Toward physical human-robot interaction control with aerial manipulators: Compliance, redundancy resolution, and input limits", 2022 International Conference on Robotics and Automation (ICRA), pp. 4855–4861, 2022, doi: 10.1109/ICRA46639.2022.9812451.

[36] G. Silano, T. Baca, R. Penicka, D. Liuzza, and M. Saska, "Power line inspection tasks with multi-aerial robot systems via signal temporal logic specifications", IEEE Robotics and Automation Letters, vol. 6, no. 2, pp. 4169–4176, 2021, doi: 10.1109/LRA.2021.3068114.

[37] C. Papaioannidis, D. Makrygiannis, I. Mademlis, and I. Pitas, "Learning fast and robust gesture recognition", in Proceedings of the EURASIP European Conference on Signal Processing (EU-SIPCO), 2021, doi: 10.23919/EUSIPCO54536.2021.9616227.

[38] V. Krátký, A. Alcántara, J. Capitán, P. Štěpán, M. Saska, and A. Ollero, "Autonomous Aerial Filming with Distributed Lighting by a Team of Unmanned Aerial Vehicles", IEEE Robotics and Automation Letters, vol. 6, no. 4, pp. 7580–7587, 2021, doi: 10.1109/LRA.2021.3098811.

[39] S. Liu, M. Watterson, K. Mohta, K. Sun, S. Bhattacharya, C. J. Taylor, and V. Kumar, "Planning dynamically feasible trajectories for quadrotors using safe flight corridors in 3-D complex environments", IEEE RA-L, vol. 2, no. 3, pp. 1688–1695, 2017, doi: 10.1109/LRA.2017.2663526.

DOI: 10.1561/9781638282839.ch9

Chapter 9

Promoting the Use of Robotics in the Inspection and Maintenance of Offshore Wind

By Andry Maykol Pinto, Aníbal Matos, João V. Amorim Marques, Daniel Filipe Campos, Maria Inês Pereira, Rafael Claro, Eeva Mikola, João Formiga, Mohammed El Mobachi, Jaap-Jan Stoker, Jonathan Prevosto, Shashank Govindaraj, David Ribas, Pere Ridao and Luca Aceto

Offshore wind energy has observed a significant growth in recent years and is expected to be the main source of energy in the future. Operations and maintenance activities are required for the complete lifetime of wind turbines, and account for up to 30% of the total operation costs. While current O&M activities are mostly man-based, making use of divers, and operators climbing the towers trough rope access, the transition towards robotic operations is expected to bring benefits in not only in terms of the costs of these operations, but also in terms of operator safety and carbon footprint of the operations. However, there are specific challenges, such as legal framework, certification of technology, and needed technological developments, that need to be tackled to allow the transition to occur.

The ATLANTIS project aims to tackle these challenges in two fronts: the creation of a testing center for robotics and the development of robotic technologies. Additionally, the project connects market needs and user's expectations of robotics from research, technology developers and system integrators, through an international consortium of ten members, from six different European countries.

Through establishing a pioneer pilot infrastructure, the ATLANTIS Test Centre, the ATLANTIS project exposes the urgency in guaranteeing early access by medium-sized enterprises (SMEs), R&D institutions or energy operators to a pilot infrastructure for validating innovative and technological solutions with high potential to leverage key market sectors and the blue growth. This pilot infrastructure will allow the demonstration of key enabling robotic technologies for inspection and maintenance of offshore wind farms. The pilot will be implemented in Viana do Castelo, Portugal, and will allow for testing, validation, and demonstration of technologies with a range of technology readiness level, in near-real/real environments.

The project also performed robotic developments that range the aerial, surface, and underwater domains, covering a total of eight different robotic vehicles (three unmanned aerial vehicles, two surface vehicles, and three underwater vehicles). Coupled with these, the project developed digital tools that aim to support and optimize inspection operations.

9.1 Introduction

Offshore wind production is one of the fastest-growing renewable energy sources and will be one of the main sources of energy in the near future, and Europe is currently the leader of this initiative [1]. The offshore wind Operations & Maintenance (O&M) global market presents a trend of growth of 17% annually, being expected to reach 11 billion Euro by 2028 [2].

Operations and maintenance (O&M) activities are required for the full operational life of a wind farm, the costs of which constitute to a significant part of the total costs of offshore wind power (up to 30% of the total cost of energy for offshore wind power [3]), with costs that can go up to 85 million euros annually, per installed GW [4] With close to 33 GW of currently installed capacity globally [5], offshore O&M costs go up to 2.8 billion year annually. Europe, with its installed capacity of close to 25 GW [5], represents the largest share (more than 2 billion euros annually), covering close to 75% of total offshore O&M costs.

The wind power sector has ambitious aims for cost-reduction to make wind power – especially offshore wind power – more cost-efficient. Most of the current O&M solutions are man-based, which have safety implications [6] and require significant resources (human resources, support vessels, etc.). A substantial contribution to cost reduction must therefore come through improved and new solutions and technologies. The use of robotic technologies in offshore wind farms can strongly contribute to these. However, there are significant challenges that need to be tackled to allow the widespread uptake of robotic technology in offshore O&M, such as the harsh environment, lack of testing, as well as legal and regulatory barriers to the introduction of these technologies in offshore environments. The ATLANTIS,[1] a European project funded under the European Union Horizon 2020 framework, aims tackle this issue, contributing to the promotion the use of robotics in offshore wind farms [7].

The ATLANTIS rationale is that robotics will be integrated and managed in the offshore wind farms within 10 years to offer substantial improvements in LCOE value by eliminating or marginalizing the use of supporting vessels for IMR operations at Offshore Wind Farms. Autonomous underwater vehicle (AUVs) and unmanned aerial vehicle (UAV) technologies have progressed to a point where it is possible to replace some of the manual inspections. Hence, the urge to enable technology testing in real offshore conditions is mandatory to enable the development of business cases for addressing concrete user requirements related to savings in commercial projects that are currently facing severe challenges related to IMR activities: vessel deployment costs, offshore logistics, reliability of in-situ repairs, condition monitoring, accessibility and availability of specialized assets, weather windows and personnel safety during offshore operations. "There are many opportunities to innovate but commercializing the best ideas take time".[2]

The remainder of this chapter is organized as follow:

- Section 9.2 will present some of the most relevant barrier to the widespread adoption of robotic technologies in the offshore wind sector that the ATLANTIS Project aims to address;
- Section 9.3 will present the ATLANTIS Test Centre, a testing facility for maritime robotics, as well as all the relevant infrastructures and services provided by the Test Centre;
- Section 9.4 will be focused on the technological developments achieved throughout ATLANTIS;
- Section 9.5 reports on the testing that has been performed for the ATLANTIS project with the platforms developed during the project;
- Section 9.6 summarizes and concludes the chapter.

---

1. https://www.atlantis-h2020.eu
2. Offshore Wind Testbeds European Experience, NYSERDA – Final Report, Report Number 18–16, July 2018.

9.2 Barriers to the Adoption of Technology

The use of robotic technologies for the inspection, maintenance, and repair (IMR) of offshore infrastructures has several potential benefits, including the reduction of Operation and Maintenance (O&M) costs, increase of operational windows, reduced downtime, improved worker safety, the scalability and replicability of the operations, among others.

However, the use of existing robotic technologies for offshore O&M in general, and offshore wind in particular, is still very limited. There are significant challenges that need to be addressed to allow an increase in the adherence of end-users and asset owners to these technologies, some of which are presented below.

9.2.1 Scientific and Technical Challenges

From a technical perspective, while capabilities of robotic systems and technologies have been subject to great advances in recent years, there are still some scientific and technical challenges that require tackling to foster their utilization in IMR operations. The harsh operational conditions of offshore wind farm environments require robotic systems to be able to deal with high winds and rough seas, with high waves and strong underwater currents, while being able to perform efficiently and reliably. These elements naturally impose challenges in terms of mobility, stability and navigation of robotic systems being deployed in these environments.

The development and implementation of advanced perception systems and algorithms is another very relevant challenges in terms of robotic capabilities for offshore operation This is particularly the case for underwater systems, where visibility decreases significantly with depth. For the case of UAV's, the limitations are related to weight, given the already limited endurance of existing systems, and to the limitations a heavy payload can impose in their maneuverability. The capability of systems performing long-term operations, or even being deployed on site permanently is appealing, in terms of resources and operations efficiency, as systems capable of on-site operation without human-presence can be a major factor in the reduction of operation costs. However, existing systems tend to have limited operation times, and are not able to perform long operations without interruptions (e.g., inspection of multiple assets sequentially). As such, developing methods to have a robotic platform to be present on site for long periods of time, allowing it to recharge itself, can be highly beneficial. Interaction capabilities, such as manipulation operations or inspections that require contact between robotic system and the turbine or underwater structure are also essential for the transition towards robotic operations. However, sea conditions and the complexity of some of the operations can severely restrict these operations, requiring further developments in stability of the robot during operation, path planning, among others.

One other aspect of robotic solutions that require further development is related to the autonomy of these systems. Autonomous capabilities can have significant impact in the outcomes of IMR operations, as they remove cognitive load from the operator during the inspection, allowing a focus on the outcomes of the operation. However, the development of fully autonomous systems requires them to be able to deal with unexpected circumstances, which can require advance decision-making algorithms. Lastly, establishing reliable and high-bandwidth communication links between the offshore robots and the onshore control center is essential for real-time data transmission, remote operation, and monitoring of the operations. However, communication in remote offshore areas can be limited or unreliable. Furthermore, underwater operations impose even more severe limitations on existing communication solutions.

While the above are just a few representative examples, they illustrate the need for technological advancements to fully realize robotic technologies as a viable solution to offshore O&M.

9.2.2 Legal and Regulatory Challenges

The deployment of robotic solutions in real offshore environments, in any recurring way, requires the compliance of these technologies with legal and regulatory requirements in place, which can limit the widespread deployment of these solutions.

Given the accelerated rate of developments in robotics, as well as the novelty applying robotic solutions to offshore IMR operations, there are a lot of unclear aspects regarding the regulations applicable to these solutions. While the use of remotely operated vehicles (ROVs) in inspection operations is already somewhat widespread, and as such there is clearer regulation for its use (NORSOK Standard U102, IMCA R018), other platforms, such as UAVs, unmanned or autonomous surface vehicles (ASVs) and AUVs are subject to more generic regulations ((EU) 2019/945, (EU) 2019/427, VTMIS Directive), that can be confusing and limit their operationality. For example, given that most regulations for vessels have been drafted with a crew in mind, they might not be directly applicable to unmanned surface vessels as these lack a crew, by definition. From a legal standpoint this can have different interpretations, including the ASV not conforming to regulations and therefore being illegal [?]. Similar examples can be found for the remaining types of platforms. In 2021, the International Maritime Organization (IMO) has completed a regulatory scoping exercise called Maritime Autonomous Surface Ships

(MASS) which has clarified a set of terms and conducted a full analysis of treaties that can be by used as a base for international regulations concerning the unmanned ships.

Additional challenges can arise from the use of cooperative systems operating in different domains (e.g., UAV taking off from an ASV). The development of these types of solutions is already underway, with a significant number of scientific developments related to taking-off and landing on moving platforms and docking of underwater vehicles into surface vehicles. However, the crossing of environments can lead to confusion on which regulation standards are to be followed. The issues presented above exemplify some of the regulation and legislation challenges that need to be tackled to allow robotic technologies to be a viable solution to offshore IMR operations. However, the tackling of these challenges is further complicated by the fast pace at which robotic technology is advancing, that cannot be accompanied by changes in legislation.[3]

9.2.3 Certification and Trust in Robotic Technologies

While the capabilities of existing and new robotic technologies have been improving rapidly, their testing and validation are severely limited, with the vast majority not being tested and demonstrated in any kind of realistic environment. Consequently, the level of trust instilled in end-users and asset owners by these technologies is not significant enough to motivate their uptake in current IMR practices. As this concern naturally stems from the lack of safe testing environments that emulate realistic conditions, tackling it can be particularly challenging. Moreover, end-users are not keen of providing access to real environments, without some degree of guarantee that the use and testing of the technology will not lead to damages to existing assets. This is particularly significant in offshore scenarios, where the environmental conditions tend to be mode adverse compared to onshore environments. While demonstrations of the operational capabilities of the technologies can be made offshore, without the use of dedicated infrastructures, this is not enough to ascertain their safety to be used in IMR operations.

In addition to this, the lack of a recognized and transversal metric that quantifies not only the benefit of the use of robotic technologies over traditional methodologies, but also their operational safety, means that the use of these technologies must be assessed case-by-case. This can require a significant amount of resources by end-users, reducing the motivation to uptake robotic technologies for their operations.

---

3. Outcome of the regulatory scoping exercise for the use of maritime autonomous surface ships (mass), imo, 2021.

9.3 The ATLANTIS Test Centre

As a way to promote not only the uptake of robotic technology for the O&M of the offshore wind sector, but also to promote and facilitate developments in the field of maritime robotics, the ATLANTIS project has developed a testing facility – the ATLANTIS Test Centre – focused on providing appropriate testing and validation of new robotic technologies. This pilot infrastructure can demonstrate key enabling robotic technologies for inspection and maintenance of offshore wind farms.

The purpose of the Test Centre is to fill the gap in terms of testing that exists between laboratory testing and validation and real offshore environments, allowing new technologies to be tested in realistic but controlled environments, and leading to increased impact of the validation of robotic technologies, all with the objective of promoting the adoption of these technologies by the OW sector. To achieve this objective, the testing, validation, and demonstration activities taking place in the Test Centre are expected to:

- Contribute to the increased end-user trust in robotic technology;
- Motivate changes in the existing legal and regulatory framework that constrict the use of these technologies;
- Incentivize robotic developments.

In this way, the ATLANTIS Test Centre will become a central piece to develop and take to market robotic technology for offshore IMR, through safe demonstrations of these technologies operating in real environments (Figure 9.1). Given the objectives and capabilities of the Test Centre, it has the potential to affect the complete robotics and wind energy value chain. The Test Centre can cater to universities and research centers, technology developers, service providers and end-users.

Figure 9.1. High level Test Centre concept.

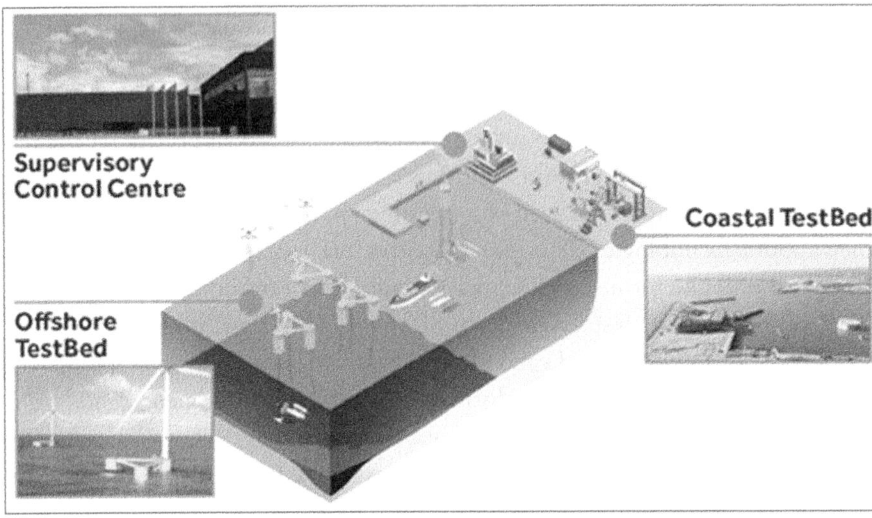

Figure 9.2. Conceptual image of the ATLANTIS Test Centre.

The ATLANTIS Test Centre is a pilot infrastructure that is geared towards the testing and demonstration of robotic technology, directed to offshore O&M. The Test Centre is capable of encompassing technologies at various levels of technology readiness level (TRL), to support and incentivize the development process. While mostly directed to technologies targeting the offshore wind sector, developments related to other applications of maritime robotics can be tested and demonstrated at the Test Centre, with the final aim of demonstrating, in a certifiable manned, the viability of robotic solutions for maritime applications.

The Test Centre is composed of two testbeds, Coastal Testbed and Offshore Testbed, and a Supervisory Control Centre (Figure 9.2). While the Coastal Testbed is focused on the de-risking of robotic technologies with lower TRL, the Offshore Testbed consists of dedicated positions within a commercial wind farm, the Wind-Float Atlantic (WFA),[4] that will be reserved for demonstrating robotic technologies in a real environment (higher TRL).

9.3.1 Coastal Testbed

The Coastal testbed of the ATLANTIS Test Centre is equipped with a floating structure that simulates an offshore floating structure of an offshore floating wind turbine. The floating structure installed is a decommissioned Catenary Anchor Leg Mooring (CALM) buoy, that provided support to the loading and discharging of liquid product cargo to/from tankers, near onshore or production fields [9].

4. https://www.edp.com/en/innovation/windfloat

Figure 9.3. Floating Structure installed on the Coastal Testbed of the ATLANTIS Test Centre (left) and point cloud reconstruction (right).

Figure 9.3 shows the floating structure, named DURIUS, as installed in the Coastal Testbed of the ATLANTIS Test Centre, in Viana do Castelo. The buoy has a diameter of 16 meters, and a height of 6 meters. The floating station is anchored in the exit of Lima River using three mooring chains. In the current installation of the buoy in the Test Centre, four meters of the structure is submerged and two meters above the water. top of the buoy, with access to the top of the floating structure enabled through two ladders, placed almost diametrically opposed in the buoy. Power and network connection is available on the buoy, being provided from shore by an underwater cable. All these elements (access stairs, mooring chains, anodes, and power cable), as well as the floating structure itself, can be used as subjects for the testing and validation of robotic technologies for offshore inspection.

9.3.2 Offshore Testbed

The Offshore Testbed is composed of the real offshore wind farm WindFloat Atlantic, containing three floating offshore wind turbines (Figure 9.4). These turbines are located between 10 and 20 kilometers of shore from Viana do Castelo. Water depths in the WindFloat wind park area range from 40 to 100 meters.

As a real environment, the infrastructures available for the demonstration of robotic technologies include all components that require inspection in a fully functional offshore wind turbine. These include:

- Wind turbine blades and tower
- Transition Piece
- Mooring lines
- Anchors
- Array cables
- Export cables.

Figure 9.4. WindFloat Atlantic.

For demonstrations related to the blades and tower or requiring close proximity to work on the floating structure, the wind turbine is required to be stopped. Additionally, access to the offshore wind park is restricted. As such, permission for the tests as well as confirmation for potential testing dates need to be obtained from the wind park owners (Ocean Winds and WindPlus).

Specific training and valid medical certification are required for all team members performing the demonstrations to access the area of the WindFloat Atlantic. The level of required training depends on the demonstrations being performed, but the Sea Survival training and OEUK (old OGUK) medical certification are a minimum requirement. Access to the Offshore Testbed is performed through a vessel departing from Viana do Castelo.

9.3.3 Supporting Infrastructures

Beyond the floating structure and respective working area, the Coastal Testbed of the ATLANTIS Test Centre has access to a set of supporting amenities and infrastructures that facilitate the realization of the tests and validations.

9.3.3.1 Infrastructures

In its installation site, in Viana do Castelo, the Coastal Testbed of the ATLANTIS Test Centre has access to all relevant infrastructures for its operation (Figure 9.5). One of the most relevant supporting infrastructures to the testing in the Coastal

Figure 9.5. Layout of the available infrastructures at the ATLANTIS Teste Centre.

Testbed is the crane that allows easy deployment and recovery of robotic vehicles into and from the water. Access ramps offer an additional deployment method for robotic systems. Additionally, there is also access to a dock and support vessels (with skipper if necessary) to monitor the operations from the water, if needed. Outside working areas are available to closely monitor the operations, while a shore control center allows for remote monitoring (see below).

9.3.3.2 Supervisory control centre

In addition to the physical infrastructures available, a Supervisory Control Centre (SCC), with a direct line of sight to the floating structure (Figure 9.6), has been implemented in the proximity of the Test Centre (see Figure 9.5). The SCC will allow for the planning, monitoring and control of the operations from shore.

The SCC has a connection network with both the Coastal Testbed floating structure and offshore Testbed (through a mobile network box), to allow for remote planning, monitoring and control of operations. All robotic vehicles performing validations or demonstrations are required to connect to the ATLANTIS Test Centre SCC, through the installation of an interoperability layer that facilitates their integration into the SCC (Figure 9.7).

In the SCC, multiple stations are available running the client used for the planning and monitoring of operations. Additionally, a small work area is available for more software intensive work (see Figure 9.8). Further details on the SCC will be presented below.

Figure 9.6. ATLANTIS Test Centre Coastal Testbed and SCC.

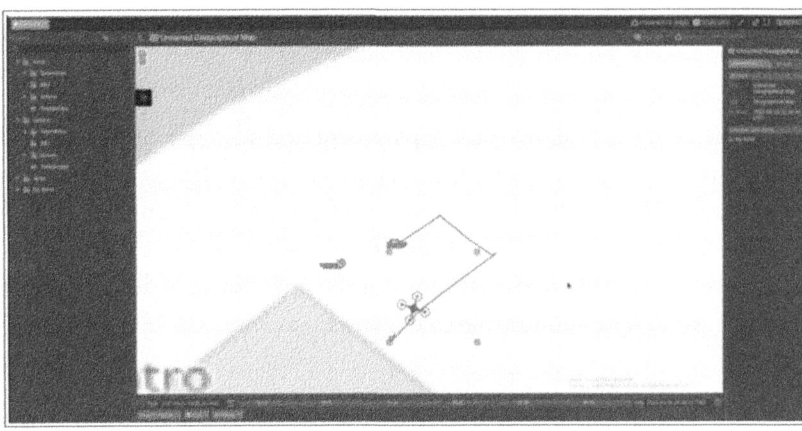

Figure 9.7. User interface of the supervisory control centre.

Figure 9.8. SCC room at Viana do Castelo.

Figure 9.9. (a) Mar Profundo Vessel (b) Back view (c) Crane (d) A-Frame.

9.3.3.3 Mar Profundo vessel

To perform the transfer between Viana do Castelo and the Offshore Testbed, the ATLANTIS Test Centre makes use of the vessel Mar Profundo (see Figure 9.9). Mar Profundo is a catamaran type vessel, with a maximum capacity of 12 passengers, including the 3-member crew. It is equipped with all required amenities to deploy robotic vehicles offshore (crane, A-frame, lifts for water access, support boat), as well as a communication link to the SCC (monitoring of operations).

9.3.4 Services and Supported Testing

The ATLANTIS Test Centre is first and foremost a facility for the testing, validation, and demonstration of maritime robotic technologies, particularly focused on the offshore wind sector. As a result, it enables users to evaluate the performance, reliability, and efficiency of the robotic solutions, providing both a safe/controlled near-real environment and a real offshore environment. Additionally, the Test Centre can provide expert technical support, including but not limited to guidance in the preparation of the operations, support troubleshooting during testing and validation operations, training in the offshore deployment of robotic platforms, connect users with relevant stakeholders, and data analysis.

In line with this, the ATLANTIS Test Centre is targeting the complete value-chain of robotics for offshore wind. As such, the services provided are directed to the needs of each of the different stakeholders. Examples of these services include:

- **Universities and Research Centers:** Field testing, technology validation, research, and development projects;
- **Technology Companies:** Technology validation and demonstration, data collection, certification of technology;
- **Service providers:** Validation and demonstration of new methodologies, assessment and de-risking of solutions, training, and certification of personnel;
- **End-users:** De-risking of technology, training of personnel, knowledge sharing and collaborations.

While offshore operations are strongly limited by weather conditions, i.e., operations are only possible during summer and early fall, testing and validation activities at the Coastal Testbed are possible year-round. All the services provided by the Test Centre include support in terms of logistics, methodology refinement and operations.

In terms of the expected costumer base for the ATLANTIS Test Centre, they include:

- Start-ups and SMEs: Smaller companies, developing innovative technologies for offshore wind energy who require testing and validation services to bring their products to market.
- Technology providers and Developers: Larger companies, that develop and commercialize robotic solutions, requiring opportunities to test and validate their products during the development stage and demonstrate them in real environments, once ready for commercialization.
- Universities and Research Centers: Institutions conducting research and development in the field of maritime robotics and/or offshore wind energy, providing them with testing facilities, technical support, and opportunities for collaboration.
- O&M Service Providers: Companies that require testing, certification, and optimization services for their equipment and systems.
- Offshore Wind Farm Owners: Companies that own offshore assets and require either training of personnel or the de-risking of their operations using new robotic solutions.

An extremely important element to promote the adoption of robotic technologies for offshore IMR is assessing and ensuring the quality of the testing, as well as the certification of the tested technology (in case of successful testing and demonstration). This process is of extreme importance, as self-certification does not provide end-users with a sufficient level of trust in the technologies to allow its deployment in place of traditional and verified methodologies. This gap can be tackled through the involvement of classification societies, that are able to devise a certification that can be recognized by third parties and end-users, guaranteeing the safety and reliability of the technology.

For this purpose, ATLANTIS has devised systematic testing procedures, included in the services provided by the center, that consider not only the robotic system's characteristics and testing objectives (defined by the Test Centre users or end-users), but also the task and associated performance indicators, to ensure the testing results are up to the highest standards. The final aim of this process is the certification of the demonstrated technology or methodology, while also ensuring all certified technology is aligned with existing safety regulations and is a viable alternative to traditional methodologies, performance-wise.

In terms of testing and validation, the ATLANTIS Test Centre is ready to support activities directly related to the scenarios defined for the ATLANTIS Project. Included in these are:

- *Scenario 1: Inspection of blades and tower*
  - Visual inspection and reconstruction using aerial vehicles
- *Scenario 2: Inspection, maintenance, and repair (IMR) of the transition piece or the floating structure*
  - Visual inspection and reconstruction of the floating structure (above water level)
  - Visual/Acoustic inspection and reconstruction of the floating structure (underwater)
  - Cathodic protection testing
  - Underwater cleaning of the floating structure
- *Scenario 3: Repair of underwater floating wind turbine cables protection systems*
  - Visual/Acoustic inspection and reconstruction of underwater cables
- *Scenario 4: Underwater monitoring over extended time periods*
  - Deployment of charging and/or docking stations
  - Visual/Acoustic underwater surveys
- *Scenario 5: Underwater close-range inspection of foundations*
  - Visual/Acoustic inspection of mooring lines
  - Visual/Acoustic inspection of anchor points
  - Cleaning of mooring lines

- *Scenario 6: Underwater monitoring of scour protection interventions*
    - Bathymetric inspection of seabed
- *Scenario 7: O&M operations supported by crewless vessels*
    - Unmanned solutions for deployment of robotic solutions (e.g., USV transporting and deploying ROV/AUV)
    - Remote control stations
- *Scenario 8: Optimization of robotic-based operations*
    - AI tools for scheduling inspection and maintenance operations
    - Predictive tools for maintenance operations

The activities presented in the list above represent the more streamlined and well-defined tests that can be performed in the ATLANTS Test Centre (either Coastal or Offshore Testbeds). However, given the resources and conditions available at the Test Centre (see previous sections), it is possible to perform a multitude of additional tests. In these situations, as the tests will not fit into the standard test list, the details need to be discussed and agreed between Test Centre coordinators and potential users.

9.4 Developments to Enable Robotic IMR of Offshore Wind Farms

While providing a testing facility for robotic technology that allows the validation and testing of robotic technologies is defined as an essential element to improve adherence to robotic technologies in the offshore wind sector, demonstrating the benefits of these solutions is another central element in the promotion of not only robotic technology, but of its development. In line with this, ATLANTIS has worked to improve existing, as well as developing new robotic technologies, to be used in offshore IMR. Coupled with these, is the development of robotic and digital tools that will impact and improve the performance of these operations. The various developments tackled by ATLANTIS were informed and guided by the previously presented scenarios, grouped into four showcases associated with specific relevant areas: turbine inspection, underwater cables inspection and maintenance, foundations, and logistics.

9.4.1 Robotic Technologies

Robotic inspection and maintenance operations in offshore environment can involve the use of robotic platforms in all relevant domains: aerial, surface, and underwater. In that sense, and in line with the showcases of the project, the developments in ATLANTIS have tackled specific challenges related to all domains, in a multitude of different vehicles.

9.4.1.1 Unmanned Aerial Vehicles (UAV)

The use of Unmanned Aerial Vehicles (UAVs) in the inspection of wind turbines has been increasing as the maturity of these solutions increases [10]. However, these require the presence of operators on site. Additionally, a significant portion of the of the operations are still performed using human-based operations (e.g., rope access). The presence of human operators on site imposes high operation costs, as well as significant safety concerns. As a result, the improvement of aerial robotic technology, particularly in remote and autonomous operation has the potential to significantly impact the IMR operations of the offshore wind sector.

In ATLANTIS, the UAV developments were focused on enabling autonomous operations and providing systems with increased operation time (endurance). With regards to the enabling of autonomous and remote operations, the project's contribution is focused on the autonomous deployment and take-off of the UAV (without human intervention) on a moving platform (which could be a supporting vessel/manned or an ASV/unmanned). The research work was centered on day/night operations – perception and navigation – robustness and reliability of the autonomous landing/take-off and resulted on the developing of a novel perception system to obtain multimodal data from the inspected offshore structures. In terms of endurance, the need for a UAV with greater endurance (between 40 and 60 minutes) and equipped with a more suitable sensor payload led to the development of new aerial systems. The new UAV (named RAVEN) satisfies the endurance requirements proposed, while also being able to transport significant payloads. Figure 9.10 presents the new platforms developed.

Figure 9.10. RAVEN UAV developed for ATLANTIS (developed by INESC TEC).

Figure 9.11. NEST landing platform (developed by INESC TEC).

To allow UAVs to be transported to and from shore without reducing their already limited operation time, a floating landing platform named NEST (Non-stationary Emersed Structure for take-off and landing) was developed. This enables the UAV to land, to be transported to and from the location of interest (as a trailer of a vessel), and to take off to perform its mission. The NEST is a catamaran-shaped platform with a landing area of 2 m × 2 m, that has a gripping surface (see Figure 9.11). This structure is composed of two 2 m × 1 m modules with detachable components that are easily unmounted and transported.

To allow the UAV to land on NEST autonomously, it has to be endowed with precise landing abilities. To allow the use of the perception system for this purpose, a three-domain and active fiducial marker has been developed, named ArTuga (Augmented Reality Tag for Unmanned vision-Guided Aircraft) [11]. This multimodal marker is inspired by the ArUco fiducial markers that are well known and widely used in the robotics field for relative pose estimation. These binary markers are coded in black and white squares, where each distinct pattern results in a unique ID. Figure 9.12 depicts the ArTuga detected by each sensor that composes the TriOPS. The relative pose from the UAV to the ArTuga is using an AI module developed based on a transformer-based architecture.

To validate these developments multiple successful tests were performed in the ATLANTIS Test Centre, Coastal Testbed, with the objective of performing a precise landing maneuver of an UAV in a floating landing platform, followed by take-off, fully autonomously using the RAVEN UAV. To detect and locate NEST landing platform, the ArTuga marker was placed in the landing platform, while the new heterogeneous perception system TriOPS was mounted on RAVEN UAV. This allowed the RAVEN to land with both accuracy and precision. Multiple successful ground tests were performed that allowed to infer the robustness and accuracy of the system. After that, multiple landings of RAVEN on NEST in water (see

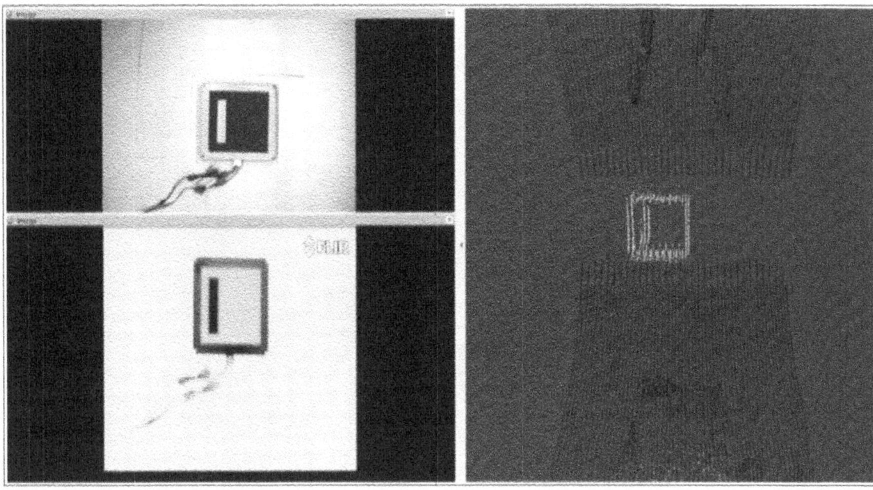

Figure 9.12. ArTuga marker detected by each TriOPS sensor.

Figure 9.13. RAVEN UAV landing autonomously on the NEST landing platform (a) and on the DURIUS (b).

Figure 9.13(a)) were successfully performed, which allowed to validate the developed system, in its intended environment. In addition to this, landing trial were performed on an platform installed on DURIUS (see Figure 9.13(b)), as to simulate the possibility of the UAV landing on the floating structure in the offshore environment.

9.4.1.2 Autonomous surface vehicles (ASV)

The use of unmanned or autonomous surface vessels (ASVs) can be one of the most relevant developments to fully remove human operators from the offshore environment, significantly impacting operational costs and operator safety. However, to safely deploy ASVs in offshore environments, there is a need to ensure their safety and reliability in terms of navigation and perception [12]. The work performed in

Figure 9.14. The designed payload (left) and the same payload integrated in SENSE (right) (developed by INESC TEC).

ATLANTIS involved the redesign of an existing vehicle's sensor payload, for both navigation and perception. The aim is to allow the vehicle to provide a better situational awareness with a wider range of applications for the inspection of offshore wind farms, such as the inspection of the scour protection integrity or the state of the structures' transition piece/splash zone. To improve navigation sensors an L1/L2 GPS/RTK and an IMU with better accuracy were included in the vehicle, while for perception, a 3D LiDAR and a stereo camera were installed at the front of the vehicle. The underwater scenario is captured through a multibeam echosounder mounted pointing downwards to retrieve bathymetric data of the seafloor, which allows estimating the seabed topography, inspect the scour protection condition and retrieve the location of known submerged structures, such as the mooring systems [13, 14]. The final version of SENSE is depicted in Figure 9.14.

To ensure the safety and completeness of the inspection tasks new algorithms were developed, namely the multi-domain mapping and the data completion method [15]. The first uses data from a multibeam and a LIDAR to create a map of the surface and underwater domains into a single representation using registration-based algorithms which allows the simultaneous inspection of the scour conditions and the transition piece.

To meet the end-user expectation regarding the offshore inspections, a new vehicle was entirely developed. This ASV was called NAUTILUS (see Figure 9.15) and has an advanced perception system that is extended by the addition of a thermal camera, as well as improvements on the previous sensors (camera and 3D LIDAR).

All modifications performed in for the surface vehicles were tested in the ATLANTIS Test Centre Coastal Testbed. The resulting reconstruction of the test center environment, using the Nautilus ASV is showed on Figure 9.16.

Figure 9.15. Nautilus ASV (developed by INESC TEC).

Figure 9.16. Reconstruction of the Test Centre performed by Nautilus during operation.

9.4.1.3 Autonomous Underwater Vehicles (AUV)

The use of Autonomous Underwater Vehicles (AUVs) for inspection operations has significant benefits, mainly in terms of costs and safety for divers, who traditionally perform the underwater inspection and maintenance operations. As such, ATLANTIS proposes a few developments for the underwater domain that aim to increase the reliability, safety, and usability of these vehicles in offshore scenarios. Of particular interest to the project were developments that related directly to the proposed scenarios: long term deployment, autonomous inspection and NDT cleaning and testing.

Long Term Deployment

To achieve the goal of the long-term deployment of an AUV directly at the offshore site of interest, an easily transportable and adjustable docking station has been proposed. A prototype of this docking station was designed and manufactured

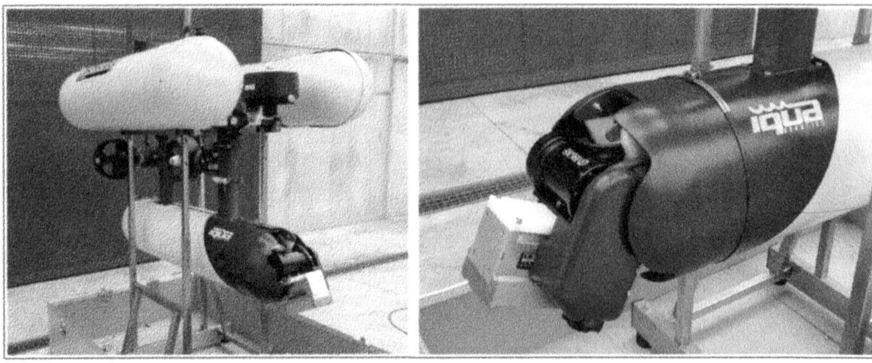

Figure 9.20. The designed optoacoustic payload (developed by IQUA Robotics).

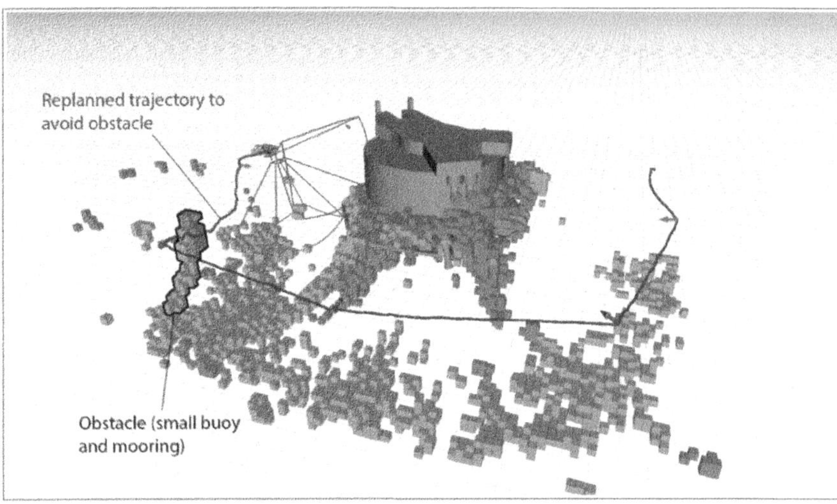

Figure 9.21. Occupancy map generated during tests taking place at the ATLANTIS Test Centre Coastal Testbed.

frame that can be adapted to both Girona AUV models (Figure 9.20). The set of cameras installed in the vehicle allows images to be acquired simultaneously from different points of view and to provide additional coverage.

The AUV, with its new optoacoustic payload was deployed and tested in the Coastal Testbed of the ATLANTIs Test Centre, as part of the project. Figure 9.21 provides a representation of the occupancy grid obtained during the inspection test.

NDT Cleaning and Inspection

To allow the expansion of the range of tasks being performed by UAVs, a complete redesign of the GIRONA 1000 payload and propulsion system was performed (Figure 9.22).

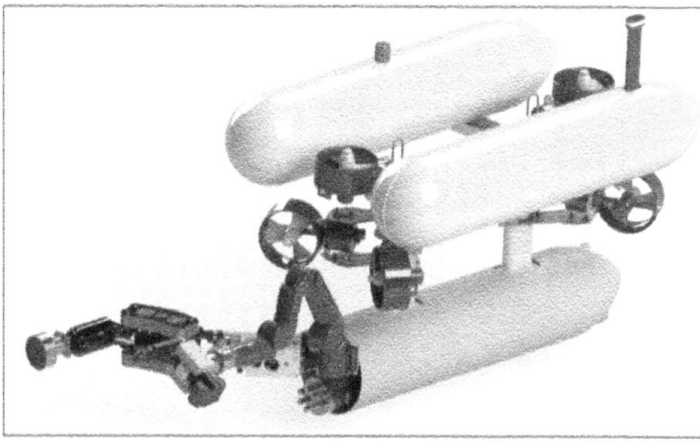

Figure 9.22. Redesigned GIRONA1000 I-AUV (developed by UdG).

The improved vehicle-manipulator system opens possibilities to perform autonomous intervention in the floating wind farm context. For the purposes of development and testing, the IMR operation that was chosen to demonstrate the capabilities of the platform is cathodic protection testing, with a contact probe. To complete this task the robot payload is composed of two manipulators, a camera, an acoustic profiler, and a laser scanner. Two manipulators are needed due to the floating nature of the inspected structure and the strong water currents that can form close to the splash zone, where one of the manipulators is equipped with a magnetic gripper, to attach to the structure underwater, while the other will be used to perform cleaning with a rotating brush, followed by a cathodic protection measurement using a CP probe. An acoustic profiler is used during the navigation phase to estimate the position of the inspected surface.

The developments performed in the scope of the project were tested in the ATLANTIS Test Centre Coastal Testbed (Figure 9.23), where the GIRONA 1000 AUV was able to successfully attach to the floating structure, while performing a cleaning operation.

9.4.1.4 Remotely operated vehicles (ROV)

The use of remotely operated vehicles (ROVs) in offshore inspection operations has become common place. However, most of the ROVs used in these inspections are limited to visual inspections of the underwater structures. As such, ATLANTIS aims to increase the ROV capabilities for operation in offshore environments in terms of TV inspection and NDT measurement, based on the Roving Bat ROV, developed by Exail. The concept for the ROV consists of a vehicle capable of navigating towards an underwater structure, landing and attaching to the structure to perform inspection and cleaning operations, while navigating on the surface of the structure using crawlers. For this purpose, the modifications undertaken include the design of a new power supply solution, installed powerful thrusters, use a digital

Figure 9.23. GIRONA 1000 in tests at the ATLANTIS Test Centre Coastal Testbed.

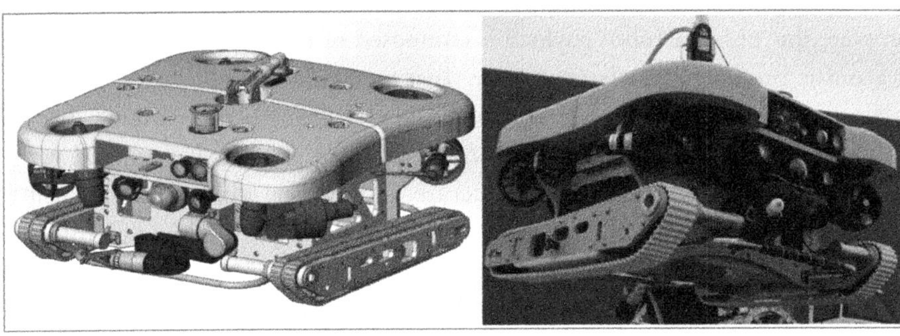

Figure 9.24. Design of the Roving Bat (left) and final system deployed (right) (developed by Exail).

control system, and adapted an agile electric arm able to position a CP Probe and a cleaning tool for anodes CP measurement. Moreover, the integration onto the Roving Bat of the MARESye camera (see below) will make possible the 3D object reconstitution in harsh underwater environments (Figure 9.24).

The improved system was tested in the ATLANTIS Test Centre Coastal Testbed, were it successfully attached to the floating structure to perform cleaning and CP measurement operations (see Figure 9.25).

9.4.2 Robotic Tools: Underwater Imaging

As mentioned above, the use of Autonomous Underwater Vehicles (AUVs) in applications such as structure inspection and repair, is rapidly gathering interest as a strong solution for these tasks [18]. Visual and acoustic inspection represent the

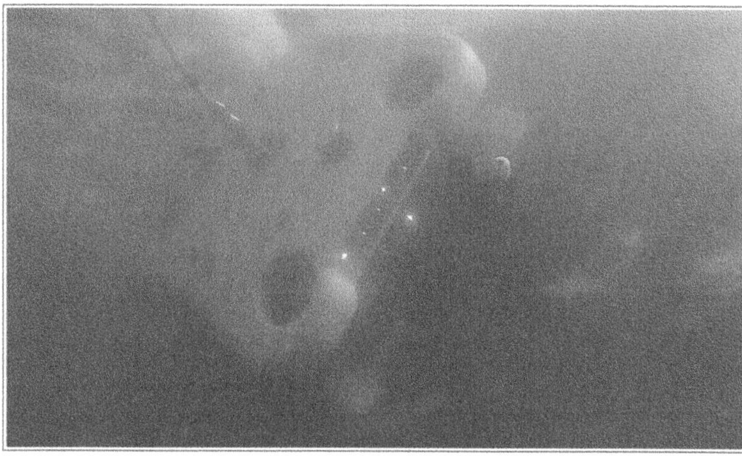

Figure 9.25. Roving Bat in operation at the ATLANTIS Test Centre.

Figure 9.26. Prototype of the MARESye (developed by INESC TEC).

main inspection methods underwater. However, according to A. Palomer *et al.* (2018) [19] one of the biggest limitations of underwater perception is the lack of quality sensors that provide 3D information. MARESye is a hybrid imaging system that combines both passive and active imaging techniques to retrieve high Signal-to-Ratio (SNR) data within these harsh sub-sea conditions. This sensor employs a high-resolution stereo setup (1080p) capable of reconstructing a dense and textured mesh of up to 150k points. Additionally, a set of laser beams can be projected onto the scene which can then be extracted to estimate sparse yet highly accurate point-clouds. This patent pending technology is engineered not only to withstand pressures of up to 300 meters in depth, but also to mitigate the effects of sub-sea phenomena. The optical system is capable of functioning in conditions where no light or texture is present – the sensor incorporates its own illumination device and pattern projection lasers.

In ATLANTIS, the MARESye system [20], suffered a complete redesign. An initial prototype was developed and built (Figure 9.26). Testing of the current

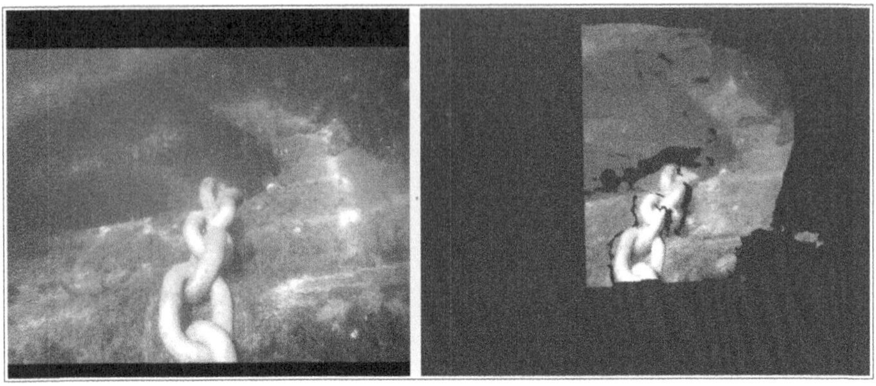

Figure 9.27. Image data captured by MARESye (left) and 3D reconstruction (right).

prototype, was performed in a hyperbaric chamber for a depth of 100 m, to ensure structural integrity and full waterproofing. This system has been validated at the ATLANTIS Test Centre Coastal Testbed, both in a standalone manner and integrated in the Roving Bat ROV, to demonstrate its versatility and the clear benefits of the integration of this imaging system in underwater vehicles. An image captured by the MARESye system, and its respective 3D reconstruction is showed in Figure 9.27.

9.4.3 Digital Tools

While the development of robotic platforms is a crucial aspect in making their deployment appealing in both terms of costs and safety, ATLANTIS has also focused on the development of digital tools that support the deployment of robotic technologies for IMR operation of offshore wind farms.

9.4.3.1 Supervisory control centre (SCC)

The planning, control and monitoring of IMR operations is essential to ensure their safety and efficiency. While for the purpose of the ATLANTIS Test Centre, the SCC is considered a physical space (as presented above), most of its technological developments are focused on a software level. As such, the SCC is a technological ecosystem with the purpose to connect the Pilot operational team to the testbed and one or more Unmanned Vehicles (UxVs) performing IMR operations. From the user perspective, a set of visualization tools have been defined that are capable of being informative about both historical and live data. To be compliant with a very high number of platforms, a web-based application that can be accessible via a modern browser was chosen. This type of technology provides out of the box the possibility to connect multiple users at once while storing the data in a centralized database. From an architecture perspective, the web application is deployed on a

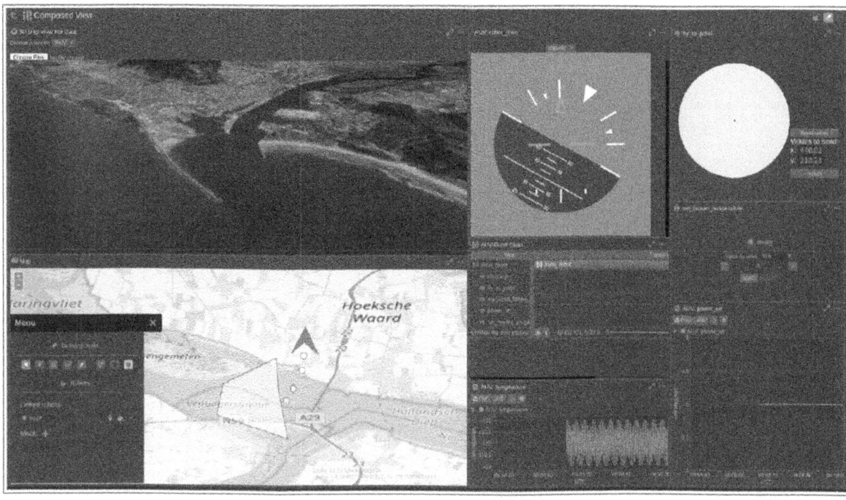

Figure 9.28. Composed View Example in the SCC User Interface Software (developed by Space Application Services).

server that provides the application itself, and a series of services that will provide the data itself. The user interface can be seen in Figure 9.28.

The communication and data exchange between the SCC and AUVs (when surfaced), UAVs and USVs in operation is done through RF. Wi-Fi and 4G are used for coastal and near-shore operations, while a satellite link is available for offshore operations. To allow an easy and smooth integration of the robotic platforms being tested into the planning and monitoring elements of the SCC, an interoperability layer was developed. This common interoperability layer is based on ROS2 and DDS and is provided to users as software. The interoperability layer has been developed in such a way that it allows for the transmission of data between the robotic platforms and the SCC using different configurations, being compatible with ROS1, ROS2 and even proprietary software on the robot's side.

9.4.3.2 Octopus

Given the dimensions of offshore wind turbines, inspection operations tend to take significant amounts of time. Additionally, planned inspection operations tend to consider multiple turbines. For this reason, the planning of the operations in an optimized way is needed to ensure minimal disruption of the operation of the turbines. As part of the ATLANTIS project, the Octopus software has been developed by ABB to support the planning of offshore operations, providing essential data regarding the operational conditions for IMR activities, especially, considering the new requirements of introducing robotic-based technology in the IMR activities. The Octopus software supplies an overview of the expected mission execution with respect to the expected weather and the specific responses of the vessel equipment,

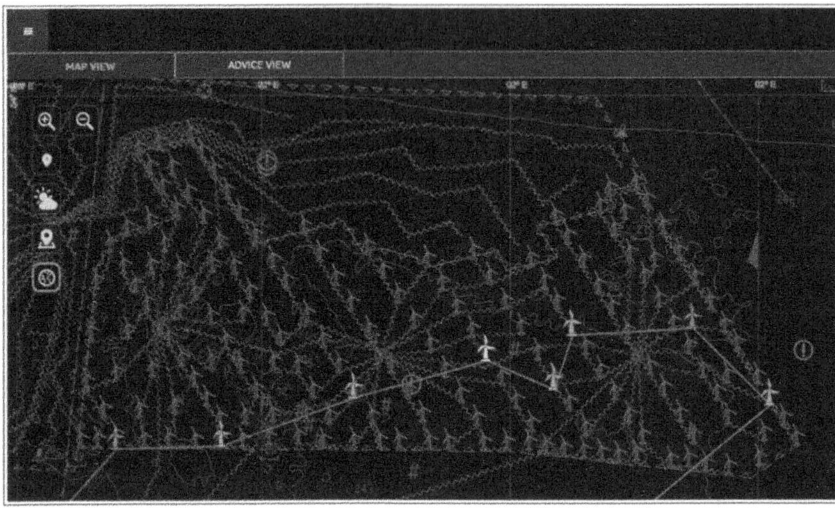

Figure 9.29. Planning and assessment map overview for windmill visits in the OCTOPUS Operation Planner.

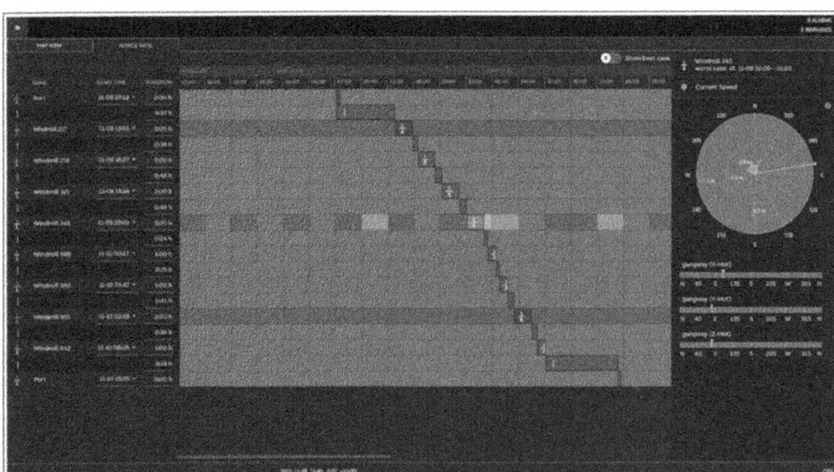

Figure 9.30. Example of a response-based project overview in the OCTOPUS operation planner.

by collecting high-resolution weather forecast for the mission and will calculating each individual stage of the mission (Figures 9.29 and 9.30). This enables the user to optimize the operation and improve the anticipation on weather standby. Thorough the use of this optimization process, ABB has identified that the use of robotic platforms, that can have less constrained operation conditions, can increase the operational windows for O&M activities by as much as 35%.

The development of the user interface design and software architecture was performed by considering industry stakeholders' inputs and industry requirements.

For the design process, the double diamond process was adapted. Design, functional requirements, and stakeholder input are prepared for rapid prototyping and demonstration with a proof of concept.

9.4.3.3 Tools for predictive maintenance

One significant concern with regards to offshore IMR is the unplanned operations, as they can cause significant disruption in the normal operation of the wind farm. As such, ATLANTIS is working to develop tools for windmill operation and maintenance (O&M) analytics, that will support the planning of IMR operations with the objective of minimizing unscheduled activities.

The ATLANTIS Test Centre Coastal Testbed is used as a case for targeted analytics development. A virtualization layer brings together the data from robots, external measurements from the Coastal Testbed and O&M analytics. To provide the data required and demonstrate the feasibility and benefits of the predictive model tools, external measurement systems have been installed in the Coastal Testbed of the ATLANTIS Test Centre. These included accelerometers, a state of motion sensor, an anemometer, a wave buoy, and cameras. The measurement system was located both on the Coastal Testbed structure, its close vicinity, and the pier (see Figure 9.31). There were three 3D vibration acceleration sensors and one state of motion sensor on the floating Testbed structure for measure its movements. Waves were monitored with a camera and a wave buoy measurement unit. Wind direction and speed were measured from the Testbed. All the measurements except the data from the wave buoy were sampled simultaneously and stored, with the system and data being collected being able to be accessed remotely.

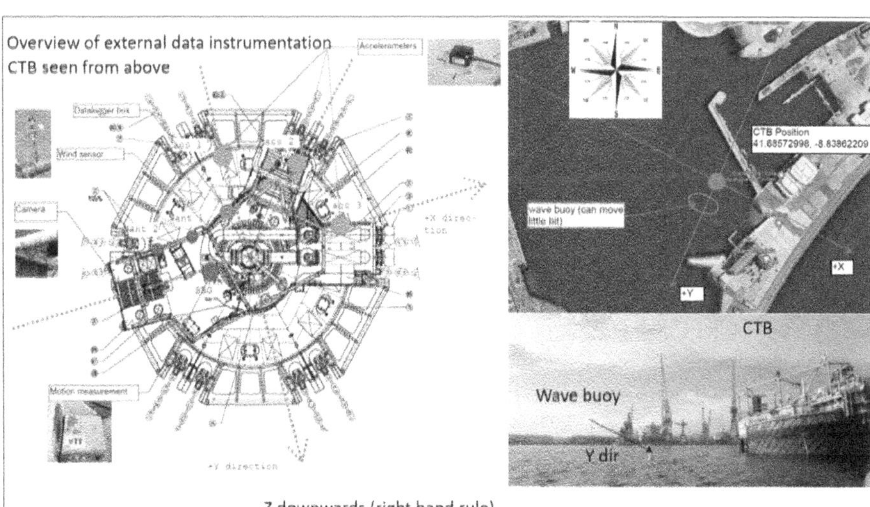

Figure 9.31. Coastal Testbed instrumentation.

Data was collected 24/7, for 5 months, and coupled with an a priori knowledge of the structure, its material properties and dimensions, a virtual representation of the Testbed was made. This virtual representation is used for estimating motions of the floating structure DURIUS at specific sea states and for evaluating highly stressed regions in the structure. Based on this data, both short (e.g., 3 hours) and long-term (whole lifetime) predictions for the loads and stresses affecting the structure can be made. Both frequency and time domain approaches were utilized. Simulations were also due to support operative criteria/limits analysis, and to estimate, based on the movements in different areas, the suitability for onboard operation and maintenance work.

9.5 Validation of the Robotic Technologies

As presented above, ATLANTIS has worked on the development of multiple robotic platforms and tools, that aim to tackle some of the technical challenges present in offshore scenarios, with the aim of promoting the uptake of robotic solutions for IMR operations in the offshore wind sector. As such, the testing, validation, and demonstration of these platforms is indispensable. Table 9.1 summarizes

Table 9.1. Summary of the testing and developments performed in ATLANTIS to date.

			ATLANTIS Testing		
Vehicle Type	Vehicle Name	Starting TRL	Scenario	Coastal Testbed	Offshore Testbed
UAV	RAVEN	–	Autonomous take-off and landing; Turbine inspection	TRL 4 → TRL 6	TRL 6 → TRL 7
ASV	SENSE	5	Inspection of the splash zone of the floating structure; Foundations and scour protection inspection	TRL 5 → TRL 6	–
	Nautilus	–		TRL 5 → TRL 6	TRL 6 → TRL 8
AUV	Girona 1000 Visual inspection	4	Autonomous underwater inspection of the floating structure	TRL 4 → TRL 6	–
	Girona 1000 Manipulation	4	Underwater cleaning and CP inspection of the floating structure	TRL 4 → TRL 6	–
	SPARUS II	4	Long term underwater monitoring	TRL 4 → TRL 6	–
ROV	Roving Bat	4	Floating structure cleaning and inspection	TRL 4 → TRL 6	–

the different robotic platforms tested, as well as their maturity, and testing scenario to the time of writing. As the project is still ongoing at the time of writing, it is expected that more of the technologies will be tested at the Offshore Testbed of the ATLANTIS Test Centre.

9.5.1 Coastal Testbed Tests

The Coastal Testbed of the ATLANTIS Test Centre presents a controlled but near-real environment, as presented in previous sections. As such, the testing of the technologies in the Coastal Testbed is aimed at validating developments and certifying the technology as ready to be deployed in a real offshore environment.

In ATLANTIS, all the developed robotic systems were tested and validated in the Coastal Testbed. With regards to the UAVs, the autonomous take-off and landing capabilities developed for the RAVEN UAV were successfully demonstrated in multiple trials, where the vehicle had to land on a floating platform (NEST) or on top of the floating structure (DURIUS) (see Figure 9.13). Additionally, tests geared towards the coordination of multiple vehicles were also conducted, with the RAVEN UAV being able to land on the NEST, while this being towed by an ASV (Figure 9.32).

With regards to the ASVs, both the SENSE and Nautilus were validated in the Coastal Testbed. The validations pertaining to the SENSE ASV were focused on autonomous navigation, in particular of its new, skill-based architecture. The tasks considered were navigation to a target point (go-to) and docking, both in an autonomous manner (Figure 9.33). For the Nautilus ASV, systems validation and testing was the focus, as the vehicle was designed and built from scratch for these tests. This included the testing of the complete system architecture, as well as the localization, navigation, and perception systems.

Figure 9.32. RAVEN being towed by an ASV to a target position after having autonomously landed.

Figure 9.33. Autonomous navigation test of the SENSE ASV.

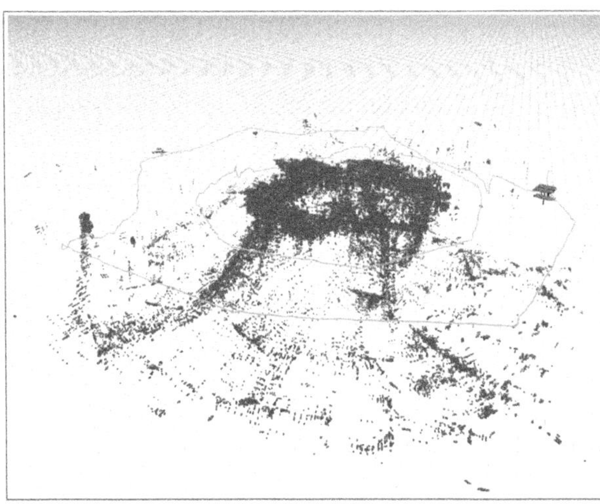

Figure 9.34. Point cloud from the acoustic sensor of the DURIUS.

The AUV tests were focused on autonomous operation. The Girona 1000 vehicle, with its optoacoustic sensor payload, aimed to complete a fully autonomous visual inspection of the floating structure. Using this payload, the inspection was divided into two different stages, with the vehicle first mapping the area for inspection at a distance using the sonar, and after completing the map of the inspection area moving close to the structure to perform a visual inspection. All of this was performed in a fully autonomous manner, While the underwater visibility was limited during these tests due to weather conditions, the vehicle was still able to generate a point cloud of the structure capable of identifying elements such as the moorings (Figure 9.34).

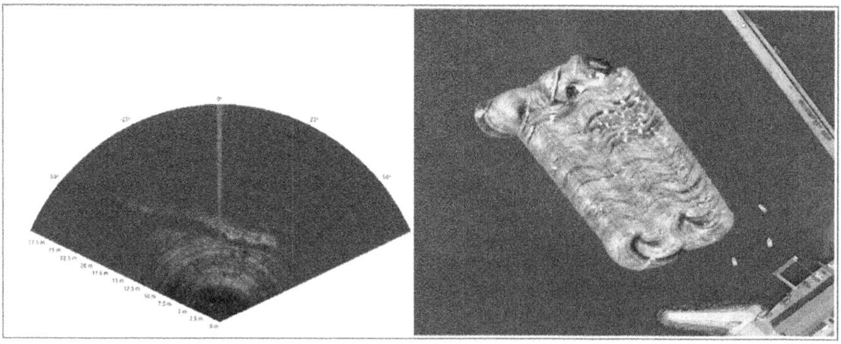

Figure 9.35. Representation of the results obtained with the forward-looking sonar.

The testing of the Girona 1000 with the dual manipulator payload was focused on autonomous manipulation and cathodic protection measurements. Using a magnetic connector as the end-effector of one of the arms, the vehicle was able to autonomously navigate towards the structure and attach itself to the floating structure. Given that the aim of these tests was to perform CP measurement test, the cleaning of the floating structure surface was needed, followed by the measurement test. Two distinct end-effectors are used for this purpose – a drill brush and a CP probe. During the tests at the Coastal Testbed, the needed cleaning was achieved in an autonomous manner, with the brus path being previously defined by the operator.

In terms of the SPARUS II AUV, the aim of the tests was not only to validate the autonomous docking process, but also to validate the autonomous navigation algorithms for underwater surveys of the seafloor. As previously mentioned, the vehicle was capable of docking, first in a remotely controlled manner and later in a fully autonomous manner. Figure 9.35 shows the results of the autonomous survey performed by the vehicle.

Lastly, with respect to the Roving Bat ROV, the objectives of the test performed started by validating the systems, as these tests were the first performed at sea. This was followed by successfully testing the attachment to the structure, as well as navigation and cleaning (see Figure 9.25).

9.5.2 Offshore Testbed Tests

With the successful completion of the Coastal Testbed validations, the next step is the deployment and demonstration of the technologies in the Offshore Testbed. At the time of writing, excluding the SENSE ASV and SPARUS II, all vehicles were tested in offshore conditions as an intermediate step before validation at the Offshore Testbed. Additionally, the RAVEN UAV and Nautilus ASV were also successfully tested in the Offshore Testbed, successfully demonstrating autonomous

Figure 9.36. UAV and Nautilus ASV in operation at Offshore Testbed.

operation capabilities in a real offshore wind farm (Figure 9.36). To the best of the author's knowledge, these vehicles (and by extension INESC TEC) were the first to be tested in a real floating offshore wind farm worldwide.

9.6 Conclusion

The ATLANTIS project aims to increase the adherence to robotic platforms for offshore O&M operations. However, there are significant challenges that need to be addressed to achieve this goal, in particular issues related to the certification and trust on robotic technologies, as well a need to further develop robotic systems to deal with the challenging conditions and tasks associated with the IMR of offshore wind farms. As such, ATLANTIS addresses both issues, through the creation of a testing center for maritime robotics, and through the development of robotic platforms and technologies.

The ATLANTIS Test center is a testing facility, the first of its kind, deployed in Viana do Castelo, Portugal, that possess the required infrastructures to test, validate and demonstrate robotic technologies and digital tools designed to be applied to maritime environments, particularly for the inspection of offshore wind farms. It is composed of two distinct testbeds, the Coastal and Offshore Testbeds, that are directed to the testing and validation of technologies at different TRL levels. The Coastal Testbed is geared towards technologies presenting a lower level of maturity (lower TRL) and supporting their development. The Offshore Testbed consists of a real offshore wind farm, the WindFloat Atlantic, containing 3 floating wind turbines, and targets technologies with a higher maturity and closer to market (higher TRL).

With the objective of tackling relevant scientific challenges, as well as shorten the gap in terms of robotic developments required to the adoption of robotic technologies in the wind sector, the ATLANTIS has worked on improving existing robotic technologies, as well as developing new tools, to meet the needs of IMR operations. The project has worked on over 7 different platforms, some of them completely new, covering all relevant domains: aerial, surface, and underwater, are focused on the development of autonomous and/or reliable capabilities in navigation, perception, and manipulation. To support these platforms, the project has also worked on areas such decision making and AI.

Beyond the physical platforms, the project has also developed a set of digital tools to support robotic IMR operations. These include the development of a Supervisory Control Centre, for remote planning, monitoring and control of the operations, the Octopus platform, for the planning and optimization of the offshore operations, and predictive maintenance models, to inform operators of the expected conditions of the assets to be inspected, and better determine the need to perform IMR operations.

While the various developments performed in the context of ATLANTIS are a significant step towards the increasing the uptake of robotic technologies, there still exists a technological and scientific gap between the current capabilities of robotic systems and operational requirements. Regardless of this, the results of the project have already demonstrated the potential benefits of the use of robotics in offshore IMR operations. From studies performed in the scope of the project, the use of robots can increase operational windows for IMR operating as much as 35%, significantly reducing downtime in unexpected maintenance operations, and reduce operation costs as much as 50% (in the case of the use of drones instead of rope access for turbine inspection). Additionally, the use of robotic platforms has the added benefit of increasing worker safety, by having robots take their place in hazardous situations (high altitudes or underwater), as well as reducing the carbon footprint of the operations, by reducing the amount of resources required for IMR operations.

As a result, it is expected that services that will be provided by the ATLANTIS Test Centre will promote further developments and facilitate the transition from man-based O&M operations to robotic-based operations, through the demonstration of the benefits the use of robotic technologies can provide. In this way, ATLANTIS contributes to the growth of the offshore wind industry, particularly in Europe, while also having the potential to promote the growth of European players in terms of technology developers and robotic service providers.

Acknowledgements

This work is funded by the European Commission under the European Union's Horizon 2020 – The EU Framework Programme for Research and Innovation 2014- 2020, under grant agreement No. 871571 (ATLANTIS).

References

[1] European Commission, "Report on the blue growth strategy. towards more sustainable growth and jobs in the blue economy," Commission staff working document, Brussels, 2017.

[2] S. Yang, "Offshore wind operations and maintenance trends 2019," [Online]. Available: https://www.woodmac.com/reports/power-markets-offshore-wind-operations-and-maintenance-trends-2019-301033. [Accessed May 2023].

[3] R. Martin, I. Lazakis, S. Barbouchi and L. Johanning, "Sensitivity analysis of offshore wind farm operation and maintenance cost and availability," *Renewable Energy*, vol. 85, pp. 1226–1236, 2016.

[4] Crown Estate, "A guide to an offshore wind farm," *Power*, vol. 2019, pp. 1–128, 2019.

[5] W. Musial, P. Spitsen, P. Duffy, P. Beiter, M. Marquis, R. Hammond and M. Shields, "Offshore wind market report: 2022 edition," National Renewable Energy Lab.(NREL), Golden, CO (United States), 2022.

[6] G+ Global Offshore Wind and Energy Institute London, "G+ global offshore wind health & safety organisation – 2017 incident data report," 2018.

[7] A. M. Pinto, J. V. A. Marques, D. F. Campos, N. Abreu, A. Matos, M. Jussi, R. Berglund, J. Halme, P. Tikka and J. Formiga, "Atlantis-the atlantic testing platform for maritime robotics," in *OCEANS 2021*, San Diego–Porto, 2021.

[8] A. Koltsidopoulos, D. Wavell and A. Gray, "Quantifying the impact of Robotics in Offshore Wind," Offshore Innovation Hub, 2021.

[9] EPCM Holdings, "SPM Calm Buoy System – The Ultimate Guide," [Online]. Available: https://epcmholdings.com/spm-calm-buoy-system. [Accessed 9 January 2023].

[10] R. M. Claro, M. I. N. F. S. Pereira and A. M. Pinto, "Energy Efficient Path Planning for 3D Aerial Inspections," *IEEE Access*, vol. 11, pp. 32152–32166, 2023.

[11] R. M. Claro, D. B. Silva and A. M. Pinto, "ArTuga: A novel multimodal fiducial marker for aerial robotics," *Robotics and Autonomous Systems*, vol. 163, 2023.

[12] D. F. Campos, A. Matos and A. M. Pinto, "An adaptive velocity obstacle avoidance algorithm for autonomous surface vehicles," in *2019 IEEE/RSJ International Conference on Intelligent Robots and Systems (IROS)*, Macau, 2019.

[13] D. F. Campos, A. Matos and A. M. Pinto, "Multi-domain mapping for offshore asset inspection using an autonomous surface vehicle," in *2020 IEEE International Conference on Autonomous Robot Systems and Competitions (ICARSC)*, 2020.

[14] D. F. Campos, A. Matos and A. M. Pinto, "Modular multi-domain aware autonomous surface vehicle for inspection," *IEEE Access*, vol. 10, pp. 113355–113375, 2022.

[15] D. F. Campos, A. Matos and A. M. Pinto, "Multi-domain inspection of off-shore wind farms using an autonomous surface vehicle," *SN Applied Sciences*, vol. 3, pp. 1–19, 2021.

[16] P. N. d. A. A. d. S. Pereira, R. D. S. G. Campilho and A. M. G. Pinto, "Application of a Design for Excellence Methodology for a Wireless Charger Housing in Underwater Environments," *Machines*, vol. 10, p. 232, 2022.

[17] J. Esteba Masjuan, P. Cieslak, N. Palomeras Rovira and P. Ridao Rodriguez, "Managed Surge Controller: A Docking Algorithm for a Non-Holonomic AUV (Sparus II) in the Presence of Ocean Currents for a Funnel-Shaped Docking Station," *Sensors*, vol. 23, p. 241, 2023.

[18] A. Sahoo, S. K. Dwivedy and P. S. Robi, "Advancements in the field of autonomous underwater vehicle," *Ocean Engineering*, vol. 181, pp. 145–160, 2019.

[19] A. Palomer, P. Ridao, D. Youakim, D. Ribas, J. Forest and Y. Petillot, "3D laser scanner for underwater manipulation," *Sensors*, vol. 18, pp. 1–18, 2018.

[20] A. M. Pinto and A. C. Matos, "MARESye: A hybrid imaging system for underwater robotic applications," *Înformation Fusion*, vol. 55, pp. 16–29, 2020.

DOI: 10.1561/9781638282839.ch10

Chapter 10

From Reality Capture to Decision-Making

By José Solís Hernández, Paula López Arévalo, Noemi Jiménez Redondo, Francisco Javier Pérez Grau, Anibal Ollero Baturone and Balázs Némedi

The OMICRON project, funded through the European Union's Horizon 2020 Research and Innovation Program, seeks to create an Intelligent Road Management Platform. This platform leverages a range of innovative technologies to enhance various aspects of the European road network, including construction, maintenance, renovation, and rehabilitation. This book narrows its focus to three key areas within the project:

1. The utilization of UAVs for reality capture through state-of-the-art technology developments.
2. The development of workflows for information management and data storage related to the Digital Twin, which stores information and maintains real-time information mirroring real-world assets.
3. Integration with the Decision Support Tool, employing various methodologies and analyses to translate captured data into actionable results, streamlining inspection, maintenance, and planning activities.

10.1 Introduction

A significant portion of European transportation infrastructures were constructed between 1960 and 1970, with a design intended for a 50-year lifespan. Assets that are past their life expectancy are the norm today, and they require extraordinary maintenance and upgrades to maintain their functionalities and to adapt to the new technical requirements and safety standards.

In this context of maturity and development across Europe, investment in transportation infrastructure at large, and particularly in road infrastructure, is primarily focused on the comprehensive management of existing road assets. In the coming years, road infrastructure investment will be further focused on road asset maintenance, in addition to ordinary maintenance tasks, major interventions will be necessary due to the fact that a large portion of the network is aged. Beyond routine road preservation tasks, the aging of a substantial portion of existing infrastructures underscores the critical need for significant rehabilitation interventions, mindful of recent events such as the collapse of the Morandi Bridge in Genoa.

The intelligent preservation of pavements and other assets holds significant potential for reducing associated CO2 emissions. Industry 4.0, marked by the rise of robotics, digitization, and artificial intelligence, and the fight against Climate Change, have reframed both the technical requisites in roadways and the methodologies required for asset preservation. The sector needs technologies for the management and preservation of road infrastructures capable of:

- Enhancing services in terms of safety and connectivity.
- Mitigating costs stemming from the substantial volume of preservation activities anticipated in the coming years.
- Increasing network capacity through intelligent utilization.
- Enabling sustainable asset management, progressing towards decarbonization in accordance with the European Green Deal (European Commission, 2019).

The OMICRON project [1], funded by the European Union's Horizon 2020 Research and Innovation Program, arises within this context with the primary objective of addressing the requirements inherent to road design, construction, and preservation processes. It orchestrates a spectrum of innovative technologies encompassing asset inspection, asset digitization, and the execution of preservation activities. Digital twin and UAVs technologies are part of the catalogue of developments included in OMCIRON's intelligent platform.

10.1.1 OMICRON's Concept

The goal of OMICRON is to develop an Intelligent Road Management Platform founded on an array of innovative technologies, aimed at enhancing the construction, maintenance, renovation, and rehabilitation of the European road network. The project encompasses the entire asset management chain, focusing on modular bridge construction, inspection digitization, predictive maintenance, and automated execution of preservation activities.

OMICRON's Intelligent Platform will facilitate the digitization and automation of numerous road management tasks that currently demand substantial labor. In this manner, OMICRON aims to pave the way towards the roads of the future, fostering Industry 4.0 and sustainability. The platform in built upon the following technologies:

1. Digitalized Inspection Technologies: The project focuses on developing technologies for inspection vehicles and UAVs to automate road inspections and enhance safety. Furthermore, these systems are connected to a V2X (vehicle-to-everything) platform to improve communication and user service.
2. Digital Twin: OMICRON's intelligent platform is based on the concept of a Digital Twin, creating a digital replica of the asset. This Digital Twin integrates diverse information sources to generate an updated and interactive representation of the road's condition.
3. Decision Support System: The OMICRON platform includes a decision support tool that works in coordination with the Digital Twin. The purpose of this system is to optimize conservation actions by calculating and predicting the infrastructure's condition using Artificial Intelligence and advanced optimization techniques.
4. Intelligent Systems for Road Construction and Preservation Processes: This section encompasses the development of (a) a modular robotic platform capable of executing various road preservation tasks, (b) a system for pavement laying, (c) Virtual and Augmented Reality systems to assist on-site personnel, and (d) a methodology for modular bridge construction.

OMICRON aims to significantly impact road management across various aspects, which can be summarized in four main areas:

- Reduction in fatal accidents resulting from conservation work, including road users and deployed personnel.
- Reduction in traffic disruptions due to conservation work.
- Reduction in regular conservation costs.
- Improvement in network capacity, measured against levels without the use of project technologies.

Table 10.1. OMICRON Use Cases.

Cluster UC	UCs	Brief description	Responsible partner
UC1. Inspection, V2X and User Support	**UC1.1**	UAVs: Management tool	University of Seville
	UC1.1.1	UAV Long range inspections	CATEC
	UC1.1.2	Multi-UAV inspection	University of Seville
	UC1.2	Terrestrial inspection vehicle	INDRA
	UC1.2.1	Innovative sensor combination	University of Cambridge
	UC1.2.2	Automatic computation of road index	INDRA
	UC1.3	V2X communications	INDRA
UC2. Routine and Emergency Maintenance Interventions	**UC2.1**	Robotic modular platform	TEKNIKER
	UC2.1.1	Installation of safety barriers	TEKNIKER
	UC2.1.2	Installation of cones	TEKNIKER
	UC2.1.3	Road assets cleaning	TEKNIKER
UC3. Extraordinary Maintenance Interventions	**UC3.1**	Signalling during construction works	CEMOSA
	UC3.2	Sealing of surface pavement cracks	PAVASAL
	UC3.3	Removal of lane markings with laser	PAVASAL
	UC3.4	Rehabilitation of surface pavement layers	EIFFAGE
UC4. Bridge Modularisation		Modular construction for bridges	Teixeira Duarte
UC5. Road Personnel Support	**UC5.1**	VR platform for robot teleoperation	LMS
	UC5.2	AR tools for worker support	LMS
UC6. Predictive Maintenance	**UC6.1**	Road digital twin	University of Cambridge
	UC6.2	Road decision support tool	CEMOSA

In this Project Use Case methodology has been applied for the definition of requirements of the variety of interconnected technologies in a structured and organized manner. After applying the methodology, six different cluster Use Cases have been defined which are composed by one or more detailed Use Cases as it is shown in Table 10.1.

10.1.2 From Reality Capture to Decision Making

From the wide variety of technologies developed within the project, this book focuses on three different aspects: (1) the use of UAVs for reality capture, (2) the developments oriented to manage the workflow of information and the data storage relative to the Digital twin, and (3) its integration with the Decision Support Tool to transform through different methodologies and analysis the captured data into results aimed to help, automate and simplify the different inspection, maintenance and planning activities. Figure 10.1 shows the schema of OMICRON's project concept, displaying its different elements and their integration on the road.

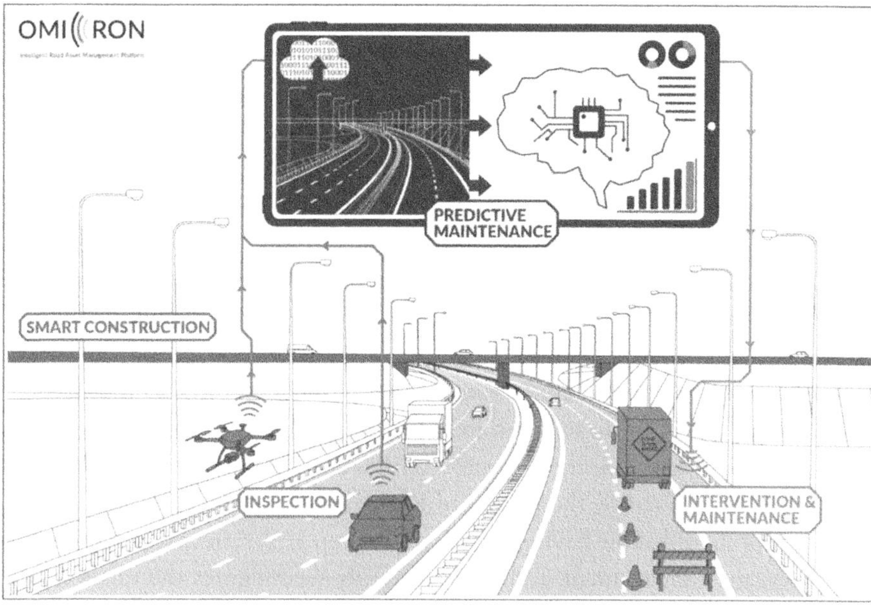

Figure 10.1. OMICRON's concept schema.

10.1.2.1 UAVs inspection technologies

The use of advanced Unmanned Aerial Vehicle (UAV) technologies can greatly contribute to the automation and digitalisation of road infrastructure maintenance actions. The objective is to apply innovative UAV technologies for the acquisition of high-quality aerial images and 3D mapping of road assets, with major attention to efficiency, safety, and reliability. The following targets are set within this project's development:

- The development of advanced road inspection modules using UAVs will allow the reduction of the exposure time by 65% of road workers and users to hazardous situations derived from inspection tasks.
- The development of digital inspection modules will allow the automation and robotisation of inspection, reducing the traffic disruption due to inspection task by at least 50%.
- The development of automated inspection will allow the enhancements in inspection and intervention tasks, reducing at least 10% the cost of road inspection and maintenance.

The effective integration of UAVs within the real-life procedures for the visual inspection of roads assets, will be highly connected to the fast deployment and the easy monitoring of their missions for an observer that must be given information two-fold. First, as concerns as the health and safety of the UAV operations, it is mandatory to get the feedback concerning the avoidance of the non-fly zones (i.e., not flying over roads/highways) and the position where the platform is (important to control that it is within an observer range that according to normative need to be involved). Second, the road asset inspector should gain insight on how the UAV is attaining the inspection of the targets. For the later, the use of a human-machine interface (henceforth referred to as **UAV Management Tool**) to be installed at a Ground Control Station (GCS) for monitoring the different flight missions is proposed. Such missions are planned assuming that the UAV will autonomously cover the desired area to acquire high-quality images, allowing for the further exploitation in OMICRON.

The development of the **UAVs management tool** is subject to the adequate guidelines and specification required to carry out UAVs inspections. The missions will involve several actors a) the inspector as End-User, b) the UAV operator and c) the Safety pilot required by the European Regulation 2019/947. The UAVs Management Tool should allow having active and backup UAVs: in case of failure of an active UAV, the Management Tool autonomously reconfigures the mission, activating a backup UAV to complete the inspection. Also, the following assumption must be taken into account:

- The inspector and the UAV operator must have knowledge of the environment and the inspection resources.
- The UAV operator, also known as Ground Control Station (GCS) operator, must have knowledge of UAV technology and performance.
- The UAVs, the UAV operator and the safety pilots are subject to current national or European regulations.

The UAVs management tool englobes visual interface, database access and communications with UAVs that allow the effective exploitation of the on-board equipment for the inspections tasks. The components of the UAVs Management Tool are presented in Figure 10.2.

The Visual Interface of the UAVs management tool is mainly aimed at providing a geographic map where the UAV position is overlapped, so as to show the space-temporal evolution of its position. The application for receiving a trajectory planning has been developed, this application's consistency is first checked against the threats/alarms published by the Aviation Authorities.

The Inspection Planning and Execution component is in charge on defining the waypoints for the flight missions which are planned according to an a-priori 3D map of the asset to inspect (and the environment around) given as input. They are Python codes which, based on ROS, allow for the correct Management for Missions with Multiple Unmanned Aerial Systems operating. These codes are embedded also in the UAV PC-companion which need to have the needed computation capacity

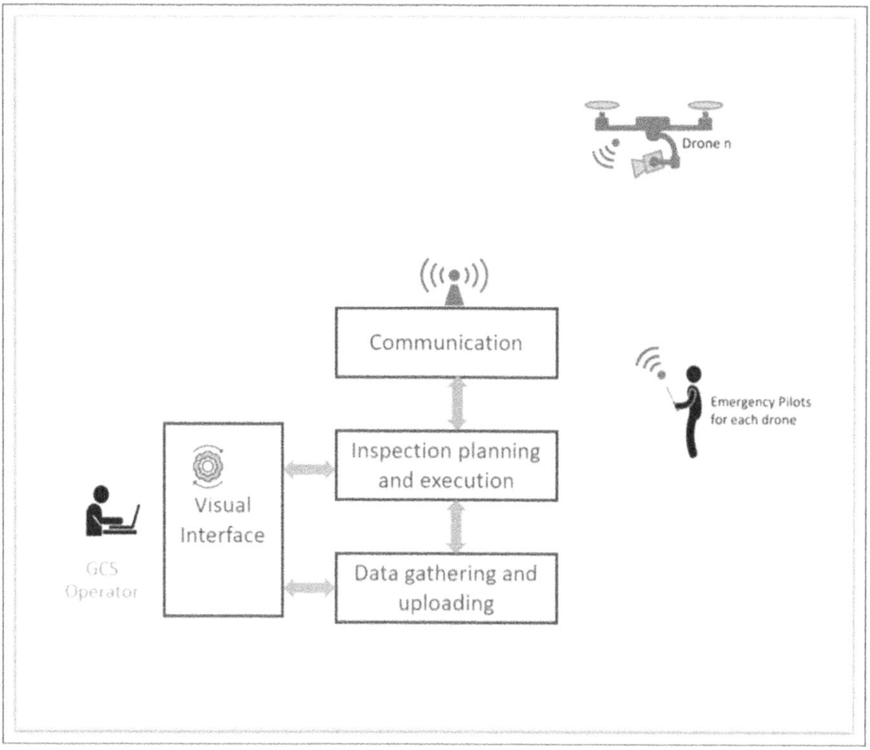

Figure 10.2. Components of the UAVs management tool.

for monitoring and controlling the different figures: GPS position, battery status, telemetry, etc.

Besides the UAVs management tool itself, two main functionalities are developed for the UAVs inspection: long range inspection (involving Detect And Avoid technologies) and multi-UAV inspections.

The **Long-range inspection** technologies will allow the inspection to be performed to cover long distances, and the Detect And Avoid (DAA) technologies will be designed and developed, based on on-board sensors and processing capabilities, that will automatically detect other manned aircraft to avoid collision threats using novel AI techniques like is shown in Figure 10.3. Following European regulation nomenclature, this type of flight will involve beyond visual line of sigh flights (BVLOS) as it is shown in Figure 10.4.

The operation of the long-range inspection is designed to proceed as follows:

- First, an inspector will make use of a UAV-based solution to examine a specific section of a road.
- After specifying the required information, the UAV operator will plan the mission accordingly.

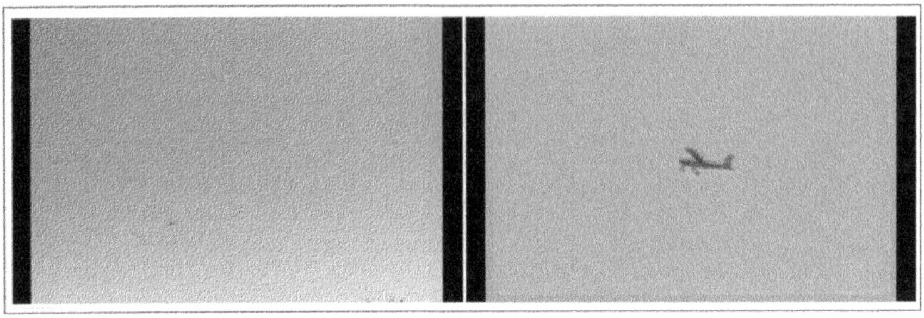

Figure 10.3. Example of aircraft detection in an image.

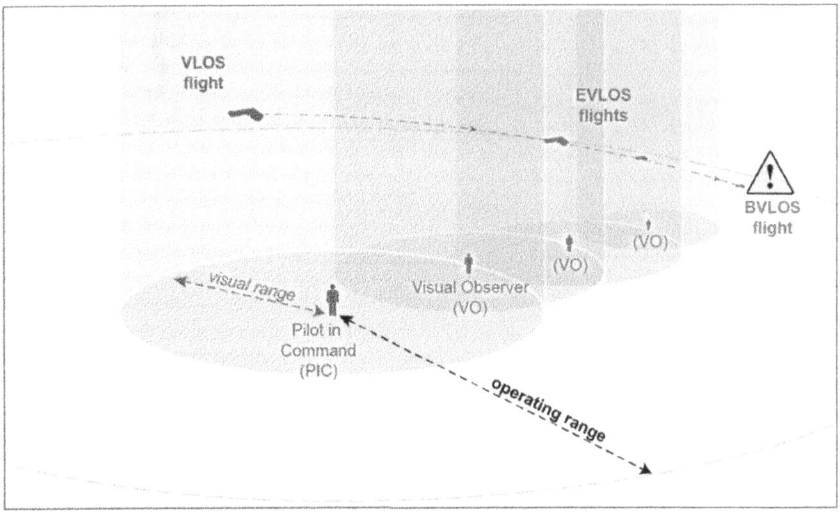

Figure 10.4. Extended range for UAV operations according to European regulations.

- The UAV will autonomously cover the desired area to acquire high-quality images for building a precise 3D map.
- Afterwards, while ensuring the safety of the full mission making use of automatic obstacle detection and avoidance capabilities.

To accomplish this operation the system includes four main subsystems, (a) the UAV platform itself, (b) the DAA system to ensure a safe flight, (c) the inspection system to acquire road data, and (d) the GCS system to monitor the operations as it is shown in the schema of Figure 10.5. The GCS module is the only one physically on the ground segment, and receives information from the rest of modules, which are integrated in the aerial segment.

The main elements of this platform are the following:

- UAV system:
    - CATEC's Platform

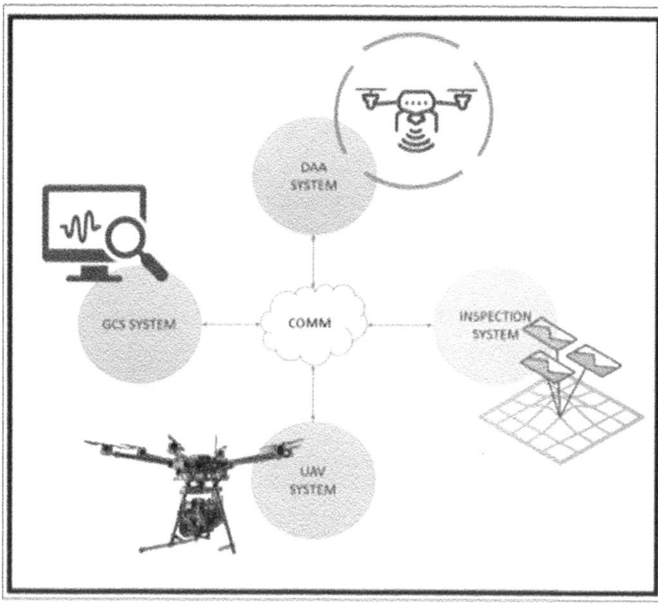

Figure 10.5. UAV-based system for long range inspections.

  - Autopilot
  - Communication system
- Inspection system
  - Camera
  - Stabilizer
  - Computer
- DAA system
  - Cameras
  - Computer

The selection of the visor sensors is particularly significant for the developments since they must be selected to fulfill the needs of the DAA while maintaining a competitive cost. The processing unit must be selected with sufficient capacity to process images in real time, so that the vision algorithms can be executed and make use of neural networks to obtain the desired results.

Once the flight mission has been carried out, the acquired images using the camera from the inspection system can be used to generate a geometric 3D model of the area of interest. To do this, this set of images must have a certain overlap that allow any specific region to be seen from different images taken at different positions and orientations. This is one of the main requirements of photogrammetry, which is used to obtain 3D measurements from 2D data.

Figure 10.6. Flight mission for test data acquisition.

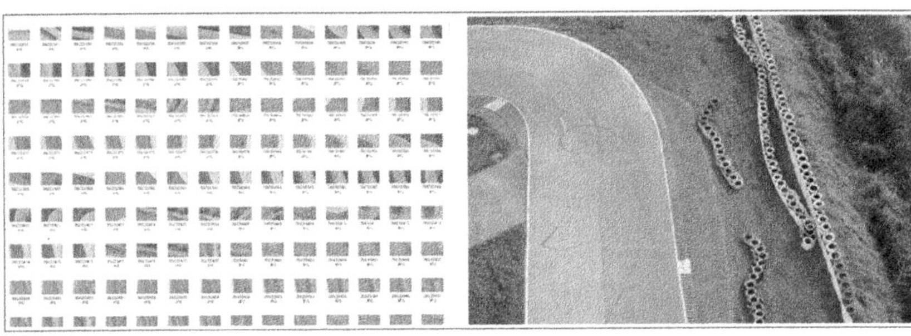

Figure 10.7. Acquired dataset (left) and example of an aerial image (right).

Figure 10.8. Reconstructed model of the inspected area.

Figure 10.6 shows a test mission planning for data acquisition which finally retrieves images as the ones shown in Figure 10.7 and allows the creation of the point cloud model shown in Figure 10.8.

On the other hand, the **Multi-UAV** system has a different motivation, the visual inspection of certain road assets such as bridges, viaducts, intersections of great urban corridors may require to the coverage a great number of targets, which potentially could necessitate of long times if they need to be covered by a solely UAV. However, these times can be shorten using multi-UAV systems: a variety of UAVs that are concurrently operated to parallelize their missions.

The system is designed to be applied when an inspector needs to make use of several UAVs deployed around a bridge/viaduct, and will allow the road asset to be inspected for retrieval of data in a more efficient manner and further, allowing for multi-views of the targeted asset. The inspector will use the UAVs Management Tool to define a multi-UAV inspection operation and the UAV operator will execute the inspection when required. There is a human-machine interface (UAVs Management Tool) where the operator monitors the concurrent flight missions, which are planned assuming that the different UAVs will autonomously cover the desired area to acquire high-quality images allowing the inspector to retrieve the data gathered after the inspection is performed.

The main objectives of this operation are listed below:

- Definition and configuration of multi-UAV inspection mission.
- Execution and monitoring of multi-UAV inspection operations.
- Flexibility to perform coordinated inspections (sharing the inspection tasks among several UAVs) or cooperative inspections (reconfiguration in case of UAV failure and/or backup UAV inclusion).
- Retrieval of data collected.

The system must assure the UAV fast deployment and operation for a flight plan provided by an operator at the UAV management tool, a fully autonomous execution of missions, the collection of information (visual cameras, with different capabilities at the different platforms) retrieved during the mission and videos stored onboard the UAVs. The main elements of this platform are the following:

- UAV system
  - Platform
  - Autopilot
  - Communications system
- Inspection system
  - Camera
  - Computer

A bridge/viaduct inspection is the kind of asset wherein the UAV technology can be used not only for reducing the hazardous to operators, but to enhanced efficiency from the use of a multi-UAV fleet. In this case, the coordination of two UAVs that are customized for gathering different views on the bridge, and whose concurrent

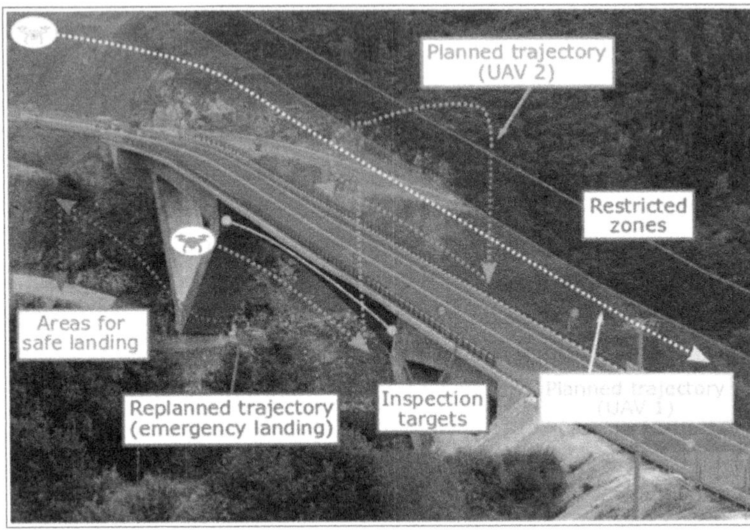

Figure 10.9. Sketch of trajectory plans.

flight missions are planned for covering all the targets (pillars, decks, and junctions, but also upper signaling and shoulder) is proposed, always with proper UAV-target distance to ensure the required GSD of the perceived images.

The acquisition of high-quality images from multiples UAVs involves the design of procedures for their respective trajectories planning for safety (typically imposing safety distance between them) but also reliability (mechanisms for collision avoidance with trees, reactivity to events, etc.). Noticeably, this means also attempting to have enough accurate trajectory tracking and positioning of the UAVs and the definition of procedures to events (event replanning for emergency landing in safety areas). Figure 10.9 shows a sketch of the rich trajectory plans that could be involved in the multi-UAV case study.

Figure 10.10 shows an example of a Multi-UAV mission carried away with two UAVs with two concurrent flight missions along with the complementary views that are finally put at the disposal of the asset.

The UAVs developments have been led by University of Seville (for the UAVs Management tool and the Multi-UAV inspection) and CATEC (for long-range inspections).

10.1.2.2 Digital twin and decision support tool

The robotization and automation of inspection and maintenance of infrastructures generates great volumes of data and new types of data. In order to exploit this, new tools and methodologies need to be developed oriented to predictive maintenance, resources optimization, enhanced management and decision-making. Digital Twin and Decision Support tools represent key elements in order to face these challenges.

Figure 10.10. Image of two UAVs simultaneously inspecting a viaduct (top). View of UAV2 (bottom left). View of UAV1 (bottom right).

A Digital Twin, following Autodesk [2] definition, is an exact digital replica of a physical asset—an up-to-date thread of information for every component of a project. It brings together design, construction, and operational data to improve efficiencies and help reduce the total cost of ownership for your customers.

OMICRON's project technologies include the development of a Digital Twin (DT) of the road infrastructure assets. The DT integrates the Decision Support Tool (DST) which uses the continuously updated information coming from the DT to implement the infrastructure condition analysis and the asset management plan optimization. DT and DST will be included in OMICRON's intelligent management tool, allowing the end-user to access the infrastructure model and information as well as to perform the available algorithms and analysis of the DST. The DT architecture is formed by different layers, the methodologies applied for the different layers definition is developed by Tekniker with the support of University of Cambridge. The DT architecture layers are:

- The physical layer
- The integration layer
- The database management layer
- The digital twin layer
- The user interface layer

The physical layer aims to collect data during the whole lifecycle of the road, from design and construction to operation and maintenance. The data is collected in different ways: inspection, monitoring, maintenance and traffic information. **The integration layer** oversees the management of static and dynamic data, data coming from the asset management databases and geometric data and semantic data from BIM models.

The database management layer defines the type of data and the repositories to store the data, based on the different sources of inspection, monitoring, maintenance and traffic data to be integrated in the Digital Twin and identified in the analysis of the OMICRON Use Cases.

The data management layer must fulfill specific data requirements. A data requirements analysis has been developed in several steps:

- Analysis of requirements. A first analysis is performed on the information exchange. This analysis is the basis for the definition of the metadata characterization template.
- Metadata characterization. A deep analysis of each of the metadata involved in the Use Cases is performed with the support of the partners leading each of the Use Cases.
- Standards analysis. Already existing and well-known standards in the domain have been analysed to evaluate their feasibility to be integrated in the data management, in the ontology model definition and in the Data Models of the repositories. IFC and CityGML have been the standards analysed.
- Legacy systems analysis. Legacy systems involved in the technical demonstrations have been analysed to understand which repositories are required for the data management.

The Digital Twin must process heterogeneous data that come from different sources and therefore have different nature and formats. Depending on these formats there are optimised technologies for their management. OMICRON aims to use the appropriate technologies to facilitate the management of each type of data while obtaining optimum performance. The types of repositories defined to store the OMICRON data are the following ones:

- For metadata which is data that provides information about the data to be acquired by the Digital Twin from the physical layer, the metadata is not the data itself, but some characteristic of the data, a semantic repository is chosen.
- Dynamic Data is information that is periodically updated over time as new information is generated, or that can be updated in a basis of a constant periodicity or may be updated at any time when a new value is generated. Time series DB is chosen for dynamic data.

- The "Asset Management Database" is intended to store data related to the operation and maintenance of the road, and some historical data that will support the implementation of the Digital Twin. Asset management is at the heart of operation and maintenance. Asset management requires a good characterization of the assets and their relationships. These assets will be subject to events and maintenance orders that influence their condition. The asset management database must be able to collect and link all data in an efficient way. The type of repository envisioned to store this data is a Relational Database.
- The Documents database is intended to store unstructured data in form of document, that is provided to the Digital Twin. The format of the document is very important to be able to access to its content and the database must facilitate the use of tools for visualizing and editing documents. When documents are edited, either internally or externally, new versions of these documents are generated, and it is essential to be able to manage this versioning. The type of repository envisioned to store these files is a Document Management System, that will allow to manage the versioning and traceability of the history of the file along the full life of the operational and maintenance of the road.
- The Static Data refers to data that is not going to change along the time or changes rarely. This could be structural data, mesh files, 2D images of the pavement, or the road, 3D images or 360 panoramic images or videos. As this is data that will not change much, it will not need advanced management and therefore specialized databases. Depending on their format and use, these instances will be stored as structured data, documents or simply files.The type of repository envisioned to store these files is a storage system, as these files will not require versioning and traceability of the history of themself along the full life of the operational and maintenance of the road. Geolocated data can be stored in specialized relational databases that provide the extension for storing files, such as for example, PostGIS that is the PostgreSQL database with the extension to store GIS files. The repository intended to store configuration data or data not in file format is a relational database.

The fundamental issue with the **Digital Twin layer** is the standards of DT product and DT process. Various data standards for data representation were considered, including IFC, CityGML, Ontology and UML to represent the specification of data. Based on the product standards, the platform can represent the DT product instances and DT product classes. In the process standards section, the specification of workflows and processing chains, using BPMN (Business Process Modeling Notation) to vissualize the Digital Twin processes.

The user interface layer provides the graphical user interface 3D and 2D model visualization and information interoperability. The platform will ensure user-friendly and convenient remote control.

The digital twin must be based on an efficient data model. With this objective a metadata characterization, an ontology, a UML (Unified Modeling Language) data model and BPMN (Business Process Model and Notation) processes have been defined.

The **metadata characterization** includes all potentially available data that will be included in the DT. Each element from the metadata characterization will be defined assigning the following fields:

- Data Identifier
- Use Case
- Inf. ID
- Description
- Digital Type
- Physical Type
- Unit
- Sample period
- Data transfer (PUSH/PULL)
- Access type.
- Read
- Write
- Read/Write
- Size
- Quantity.
- Data Provider.
- Data Responsible.
- Metadata Provider.
- Repository.
- Integration Protocol.

For the DT an **ontology** conceptual representation for roads is elaborated. An ontology is a formal description of knowledge as a set of concepts within a domain and the relationships that hold between them. Ontology does not only introduce a sharable and reusable knowledge representation but can also add new knowledge about the road domain. To enable such a description, OMICRON needs to formally specify components such as individuals (instances of objects), classes, attributes, and relations as well as restrictions, rules, and axioms. As one of the building blocks of Semantic Technology, ontologies are part of the W3C standards stack for the Semantic Web. They provide users with the necessary structure to link one piece of information to other pieces of information on the Web of Linked Data. Because they are used to specify common modelling representations of data from distributed and heterogeneous systems and databases, ontologies enable database interoperability, cross-database search, and smooth knowledge management.

The creation of an ontology in OMICRON is intended to improve data management. Some of the major characteristics of ontologies are that they ensure a common understanding of information and that they make explicit domain assumptions. As a result, the interconnectedness and interoperability of the model make it invaluable for addressing the challenges of accessing and querying data in large organizations. Also, by improving metadata and provenance, and thus allowing organizations to make better sense of their data, ontologies enhance data quality.

To develop the ontology the first step is the identification of the road assts, elements and relations. Ontologies must be "carefully designed and implemented" to properly model all the necessary information for their final use. Also, ontology development is becoming more and more centred in reuse. Given the above, it is important to follow a well-defined design methodology to develop ontologies that are optimal both for their intended function and to be reused by others. In the development process of the OMICRON ontology, the selected methodology is LOT (Linked Open Terms), in its industrial version. For OMICRON's ontology Use Cases as well as technical and main demonstrators' requirements will be considered to develop the competency questions for the ontology elaboration.

At this starting point, some of the main assets at road level identified are road, bridge, pavement, signals, lighting/systems, safety barriers and marking lines. And some of the main assets involved in the operational and maintenance activities are the following ones: UAV, robot, safety cones, camera, laser for painting removal, sealing tool, inspection Vehicle, V2X, sensors, RSU (Road Side Units).

After the creation of the metadata characterization and the ontology, **UML data models**. UML is a general-purpose visual modeling language that is intended to provide a standard way to visualize the design of a system. OMICRON UML data models are created for the OMICRON repositories, these data model's creation based on IFC5 and CityGML data models is comprehensive to represent the data structure, including asset tree and KPI, measurements, events, alarms, maintenance management, etc. This UML data model can be used for the OMICRON project maintenance management. In the practical process, the KPI will be extended to meet the real requirement in the OMICRON project. Asset Tree and KPIs are defined based on the DT product. Measurements, events and alarms, maintenance management are defined based on the DT process.

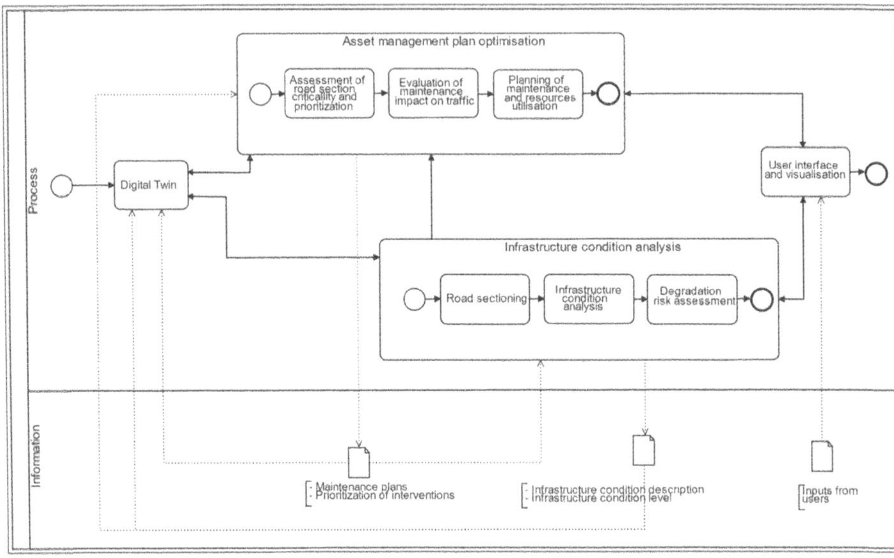

Figure 10.11. BPMN model for UC6.2. decision support tool.

BPMN models are created providing a general overview and objectives for each OMICRON Use Case by modelling each of them with explicit reference to the inputs, outputs and processes involved. To illustrate this, Figure 10.11 shows the BPMN model for the UC6.2 (Decision Support Tool).

Once all the information relative to the Digital Twin is specified and the data models are generated, implementation must be carried out. First, the digital twin model must be created implementing the information with a BIM-GIS approach (with Prointec in charge of the BIM-GIS integration) and second, the DT updating must be orchestrated considering new data from miscellaneous sources, the connection to dynamic data, the continuous update of the DT model about the state of the infrastructure and the query retrieval from the DT.

As it was previously stated, the DT information is the base for the DST developments. Once the end-user, in this case the road concessionaire accesses the DST, different analysis will be available for its execution. The execution of the different analysis will provide user-oriented information about infrastructure condition, the road concessionaire will implement this information on the decision-making process to elaborate the final work orders.

The DST is being implemented to improve predictive maintenance through the optimization of planning interventions and resources, the DST algorithms can be categorized in two differentiated modules: (1) The DSS condition analysis which is the module in charge of the infrastructure status evaluation, degradation assessment and condition prediction and (2) the DSS optimization which is the module responsible for the maintenance planning functionalities, supporting the asset management, the decision-making and improving availability and reliability. These modules include:

- Pavement state prediction: analysis of historical information from IRI, SFC, deflections, etc. to assess road condition, application of AI techniques for prediction and use of results to apply decision-making methodologies. These tasks include data analysis and data mining techniques as well as GIS management methodologies, and are led by CEMOSA.
- Structural Health Monitoring (SHM): analysis of IoT data coming from monitored bridges to analyze infrastructure condition. One of the approaches for the infrastructure assessment is the Operational Modal Analysis for Structural Health Monitoring for dynamic characterization. Depending on the quality and quantity of available data coming from the IoT sensing system and after its analysis regarding data management, data mining, feature selection and others, different possible models can be applied (from statistical process control and outlier analysis to supervised learning or model updating) to detect, localize, classify or even predict damage. The partner in charge of SHM developments is CEMOSA.
- Enhanced Maintenance Planning: University of Genova is the main responsible of the improved maintenance planning. Information coming from infrastructure condition and operational aspects is combines to generate maintenance plans. A core element from this development is the traffic simulation.

On the visualization layer side, a user-friendly platform in being created for the road operator to access all information relative to the DT and DST. The platform, which development is led by Regens, will be accessible by different users with different licenses using the correspondent credentials. The licenses are:

- Viewer: they can only access the data in viewer mode.
- Administrator: they can view the data and modify facility information.
- General Administrator: they can view and modify the data as well as adding/editing facilities and users to the platform.

The platform will include all the infrastructure assess and a specific visualization if generated for the Generator Administrator to add or remove facilities.

The connection between the Digital Twin and the dynamic data sources is being developed for it to be automatically refreshed inside the system while the Information provided by static data sources and punctual inspections performed by traditional means or OMICRON's inspection technologies will be added to the

DT using their correspondent platform menus by the administrator or the general administrator.

Any user with access to the platform will have access to the different visualization menus from the platform that can be divided into:

- Digital Twin menus allow the access of all useful information stored in the Digital Twin, this includes:
  - Asset menu allows the visualization of the GIS and BIM models of the digital twin. It is also the "home" visualization of the platform.
  - Data menu's purpose is to provide access to the digital twin database, which has been categorized into:
    - Historic results menu stores the historical data of the system and other static information.
    - Interventions menu stores information related to maintenance operations carried out by OMICRON technologies. This information is referred to location and date. GIS information and visor is maintained in these menus.
  - Monitoring menu stores information coming from dynamic sources such as IoT sensors.
  - Inspection results menu includes the results of inspections carried out by OMICRON technologies. There is information relative to:
    - UAVs: it provides UAVs inspection results. BIM-GIS visor is also accessible.
    - Terrestrial: it provides terrestrial vehicle results. BIM-GIS visor is also accessible.
    - Point-cloud: this visualization shows the point cloud of the assets.
    - Auscultation: results from auscultation analysis (videos and results).
  - Analytics menu shows the results related to intermediate data processing carried out by the digital twin, such as aggregate tables of road parameters.
  - DSS analytics menu shows past analysis performed applying Decision Support Tools which were stored to be consulted.
- Decision Support Tool menus allow users to run the DST analysis on real time.

Figure 10.12 shows some examples of a preliminary version of OMICRON's platform. On the top image the point-cloud visualization menu is shown, while on the bottom image a list of assets registered in the platform is shown in the assets menu. Both menus show on the left the control to access other menus and drop-downs to select elements which complement the manual selection of the interface.

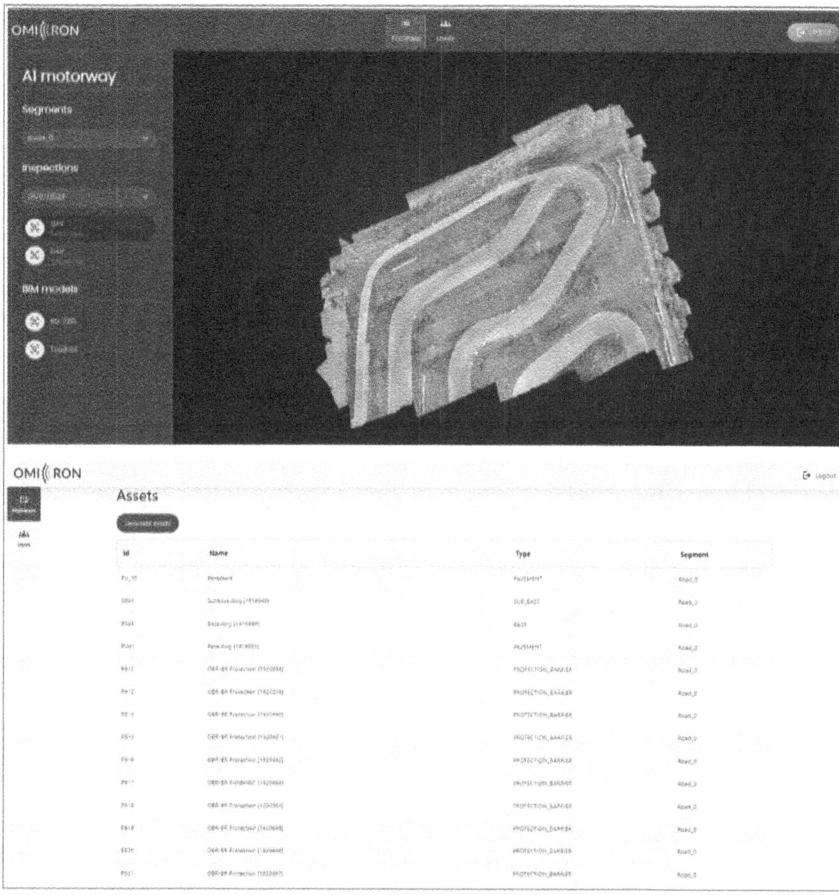

Figure 10.12. OMICRON's platform menu example. Top image: Point-cloud visualization menu. Bottom image: List of assets from Assets management menu.

10.1.3 Conclusion

OMICRON's project emerges on a context of increasing need of aged infrastructure surveillance, maintenance and rehabilitation, also being highly influenced by the worldwide shared objective of fighting climate change. OMICRON aims to enhance safety, connectivity and network capacity while mitigates costs and emissions through the development of an Intelligent Road Management Platform based on an array of innovative technologies focused on construction, inspection, digitalization, maintenance and preservation. The platform is built upon:

- Digitalized inspection technologies
- Digital Twin
- Decision Support System
- Intelligent System for Road construction and Preservation Processes

Different UAVs technologies are developed as part of the digitalized inspection technologies:

- UAVs management tool: human-machine interface for monitoring the UAV's inspection missions. Its visual interface will show a map with the UAV's position evolution with time and will be prepared to receive the mission's trajectory.
- Long-range inspection technologies will allow the inspection to cover long distances, including BVLOS flights and it includes a Detect and Avoid system.
- Multi-UAV system: this development allows complex mission to be executed using the simultaneous operation of several UAVs.

Another key element of the Intelligent platform aligned with the scope of the book is the Digital Twin and its Decision Support System. The DT stores information and mirrors real-life assets on real time. For its implementation its architecture is defined creating different layers. Also, different elements for the data analysis, data model and methodologies are defined:

- Metadata characterization
- Ontology
- UML data models
- BPMN process models

The Decision Support Tool is based on the DT continuously updated databases and allows its users to perform different analysis to help on the decision-making process relative to structural health monitoring, pavement state prediction and enhanced maintenance planning.

UAVs technology, Digital Twin, Decision Support tool and the rest of the technologies from the different Use Cases will be integrated into the Intelligent platform tool, this platform allows end-users to access and analyse all the information available within the project in an accessible and user-friendly way.

Acknowledgements

This project has received funding from the European Union's Horizon 2020 Research and Innovation Program under Grant Agreement No. 955269, signed between the following parties: CEMOSA, Tekniker, AISCAT, Autostrade per l'Italia, CATEC, University of Seville, University of Cambridge, Panepistimio Patron (LMS), University of Genoa, Eiffage Infraestructuras, Pavasal, Indra, Regens, ESCI, Armando Rito Engenharia, Teixeira Duarte, Autostrade per l'Italia and CINEA, with delegated powers from the European Commission.

Specifically, the group of organizations which have participated in the UAVs' developments are CATEC and University of Seville, while in the Digital Twin's developments the partners where CEMOSA, University of Cambridge, Tekniker, Regens, Prointec and University of Genova.

References

[1] *H2020, OMICRON. Intelligent Road Asset Management Platform*, 2021.

[2] Autodesk. Un gemelo digital. [Online]. Available: https://latinoamerica.autodesk.com/campaigns/digital-twin.

DOI: 10.1561/9781638282839.ch11

Chapter 11

Robotics-Enabled Roadwork Maintenance and Upgrading

By Nikolaos Bakalos, Iason Katsamenis, Eftychios Protopapadakis, Carlos Martín-Portugués Montoliu, Yannis Handanos, Franziska Schmidt, Olov Andersson, Helen Oleynikova, Miquel Cantero, Ilias Gkotsis, Elena Avatangelou, Nikos Frangakis, Dimitrios Bilionis, Dimitris Vamvatsikos and Maruša Benkic

In the contemporary era, road infrastructure stands as a paramount public asset, serving as a linchpin for economic expansion and societal progress by facilitating access to essential services such as education, health, and employment opportunities. These complex networks, critical for sustaining modern economies, necessitate meticulous upkeep and modernization efforts to ensure the safety, health, and efficiency expected by road users. This chapter delves into the strides undertaken by the EU-funded HERON project, aimed at transforming the maintenance landscape of road infrastructure through the implementation of an advanced integrated system. This project heralds a significant evolution in road management strategies, potentially lowering the frequency of accidents, diminishing maintenance expenditure, and amplifying the capacity and efficiency of the road network at large. Central to the HERON project is the advent of a sophisticated autonomous robotic ground vehicle, augmented with the support of independent drones. These technological advancements work in synergy, utilizing high-resolution sensors and

3D mapping scanners to bring a new dimension of accuracy and foresight to road maintenance protocols. Moreover, the integration of artificial intelligence toolkits serves to streamline and coordinate maintenance and upgrade workflows, thereby heralding a new era of efficiency and safety in road infrastructure management. This chapter provides a comprehensive overview of the HERON project, exploring its potential to redefine road infrastructure maintenance and usher in an epoch of prosperity and well-functioning economies, sustained by robust and well-maintained road networks.

11.1 Introduction

In the burgeoning field of infrastructural technology, the synergy between robotic platforms and computer vision is reshaping the paradigms of road infrastructure monitoring. This integration heralds a new era of scientific innovation and societal progress, catering to the dual objectives of efficiency and safety. The following sections will delve deeper into the nuances of this transformation, setting the stage for an insightful exploration in subsequent chapters of this book.

To begin with, the scientific impacts of these advancements cannot be overstated. Not only do they promise unprecedented automation in data collection, but they also facilitate real-time monitoring and analysis of road conditions, a leap forward from traditional methodologies. These technologies have transcended barriers, enabling remote sensing capabilities to probe into areas previously deemed inaccessible, thus broadening the horizon for comprehensive infrastructure assessment. This constitutes a significant leap in scientific innovation, paving the way for data-driven insights and decision-making processes that are grounded in precision and accuracy.

From a societal perspective, these technological strides offer a plethora of benefits, enhancing the safety and efficiency of road monitoring systems. Robotic platforms can navigate hazardous areas or high-traffic zones with ease, thereby mitigating risks to human personnel. This advancement not only ensures the well-being of the workforce but also augments the pace of response to infrastructural issues, thereby fostering a safer and more streamlined road network for the general populace.

Furthermore, the amalgamation of these technologies with existing systems such as Geographic Information Systems (GIS) promises an integrated approach to urban planning and traffic management. This collaborative approach is likely to facilitate informed decision-making, thereby contributing to the development of urban landscapes that are well-orchestrated and conducive to smooth traffic flow, a change that has far-reaching implications on community satisfaction and overall societal well-being.

The advent of predictive maintenance, empowered by machine learning algorithms, denotes a significant shift in infrastructure management strategies. This proactive approach facilitates timely interventions, optimizing resource allocation and reducing the probability of escalated damages, thus ensuring a smoother, more reliable road network for society. Moreover, the automated reporting systems integrated into these platforms expedite the response times to emerging issues, offering a direct societal benefit in the form of efficient traffic management and safer road conditions.

At the economic frontier, these approaches are sculpting a cost-effective path for road infrastructure maintenance, reducing the dependency on labor-intensive practices. This evolution promises not only financial savings but also an optimized maintenance schedule, circumventing the necessity for expensive emergency repairs and fostering economic prudence in resource management.

Finally, in the realm of research and development, these technologies serve as a fulcrum for further innovation. The data amassed forms a rich repository for researching road wear patterns, facilitating the development of robust materials and construction techniques. Additionally, these data points are critical in developing simulations and models that test new road materials and designs, nurturing an environment of continual growth and innovation in road construction.

11.1.1 Previous Approaches

HERON stands as one of the initiatives funded by the European Union under the Horizon 2020 program, specifically focusing on the maintenance of road infrastructure. Within HERON's scope is the implementation of a road infrastructure blueprint developed by affiliated projects. The primary objective of HERON is to pioneer advanced engineering solutions for interconnecting and facilitating seamless transitions between different transportation modes in the event of severe disruptions affecting one mode of transportation [1].

Numerous projects, similar to HERON, have received funding from the European Union to create an autonomous ground robotic vehicle, complemented by autonomous drones. This initiative is intended to enhance the monitoring, assessment, and maintenance of road infrastructures. One such EU-funded project, known as InfraROB [2], is dedicated to the automation, robotization, and modularization of road construction and maintenance tasks. More specifically, it will design autonomous robotized equipment and machinery for tasks such as line marking, repaving, and the repair of cracks and potholes. Additionally, collaborative robotized safety systems will be developed to ensure the safety of both construction workers and road users. The project also seeks to integrate pavement management and traffic management systems, aiming for a comprehensive, unified approach to the management of road infrastructure and real-time traffic [3].

In parallel, the OMICRON project [4], which is financially supported by the European Union, aims to create an intelligent asset management platform (IAMP) equipped with a diverse array of region-specific cutting-edge technologies. These technologies are intended to enhance the construction, maintenance, renewal, and enhancement of the European Union's road network. The project will encompass the entire road network system and will specifically involve the implementation of digital inspection technologies, the development of a road digital twin, the creation of a decision support tool, advancements in intelligent construction, and the provision of intervention solutions for infrastructure-related issues. To facilitate this, the IAMP will be seamlessly integrated with a digital twin built around the principles of Building Information Modeling (BIM). This integration will enable the industrialization and automation of various road management functions and will be demonstrated in Italy and Spain.

Finally, the PANOPTIS project [5] has set its sights on enhancing the resilience of road infrastructures and ensuring their continued functionality in adverse conditions, such as extreme weather events, landslides, and earthquakes. The project's primary goal is to amalgamate downscaled climate change scenarios tailored for road infrastructures with simulation tools encompassing structural and geotechnical aspects, as well as real-time data collected from both existing and innovative sensors. This amalgamation will result in an integrated tool designed to empower operators in the more effective management of their infrastructures across the planning, maintenance, and operational phases.

All of the previously mentioned projects, including HERON, are recipients of funding from the European Union's Horizon 2020 Research and Innovation Programme. It's worth noting that HERON represents one of the three Research and Innovation Actions supported within the scope of ameliorating environmental impacts and achieving fully automated infrastructure upgrades and maintenance under the Horizon 2020 program SOCIETAL CHALLENGES – Smart, Green and Integrated Transport.

11.1.1.1 The HERON contribution

The HERON project stands as a beacon of innovation, setting its sights on revolutionizing the process of maintaining and upgrading road networks. At its core, the initiative seeks to create an integrated automated system capable of handling a range of roadworks tasks. These encompass a plethora of activities including sealing cracks, patching potholes, rejuvenating asphalt, autonomously replacing CUD elements, and refreshing road markings. Moreover, it extends its functionality to support both the pre and post-intervention phases, which notably involves conducting visual inspections as well as automated and controlled deployment and retrieval of traffic cones.

Delving into the specifics, the HERON system embodies a multi-faceted approach. Firstly, it incorporates an autonomous ground robotic vehicle, which works in harmony with supportive drones to meticulously coordinate both maintenance works and the necessary procedures before and after the intervention phase. This vehicle serves as a hub, housing a diverse range of robotic equipment furnished with an array of sensors and actuators. These tools are adept at executing tasks such as cutting and filling, placing surface materials and compacting them, installing modular components, and undertaking 3D mapping through laser scanners.

Furthermore, an advanced sensing interface finds its place both on the robotic platform and within the Road Infrastructures (RI), enhancing the monitoring capabilities. This interface grants a heightened situational awareness, offering a detailed insight into the structural nuances and functional conditions of the RI, along with scrutinizing road markings.

The operational heart of the system is a sophisticated control software, which seamlessly bridges the sensing interface with the active robotic equipment, thereby orchestrating a synchronized performance. This software is complemented by Augmented Reality (AR) visualization tools, which grant the robotic system the ability to discern surface defects and scrutinize road markings with an unprecedented level of detail.

In its pursuit to streamline operations, HERON employs AI-based toolkits that function as a middleware, serving dual pivotal roles. Firstly, these toolkits are entrusted with the task of optimally coordinating road maintenance and upgrading workflows, ensuring a seamless operational cadence. Secondly, they are responsible for the intelligent processing of data harvested from both the vehicle and infrastructure sensors. This data processing is crucial in guaranteeing safe operations while avoiding disruptions to regular traffic flows and other routine operations.

Central to the functionality of HERON is an enhanced visualization user interface which integrates all collected data, thereby offering a comprehensive platform that facilitates informed decision-making. Furthermore, communication modules are intricately woven into the system, fostering Vehicle-to-Infrastructure or Vehicle-to-Everything (V2I/X) data exchanges. These exchanges are fundamental in not only foreseeing maintenance requirements but also augmenting user safety.

HERON emerges as a pioneering venture, promising a modular design that can adapt to various transport infrastructures. By doing so, it harbors the potential to significantly diminish fatal accidents and maintenance costs while alleviating traffic disruptions. Ultimately, this leads to an uptick in network capacity and efficiency, marking a substantial step forward in the domain of road infrastructure maintenance and management.

The rest of this chapter is structured in the following way. Section 11.2 presents the project demonstration sites and discusses the application scenarios. Sections 11.3–11.5 present the HERON technological components, starting with the developed computer vision toolkits (Section 11.3), the robotics platform (Section 11.4) and the project's components responsible for visualization and increasing the overall situational awareness (Section 11.5). Section 11.6 presents the chapter's conclusions.

11.2 Application Scenarios

11.2.1 Demonstration Sites

HERON will deploy the technological innovations described in Sections 11.2–11.4 in three different demo sites, in France, Greece, and Spain.

11.2.1.1 Greek demo site: Olympia Odos

Olympia Odos is a Motorway Concession Project of particular strategic importance on national and regional levels for the development of the Peloponnese and Western Greece, as it connects Athens with North Peloponnese, Western Greece and the Port of Patras.

The Motorway is 202 km in length and comprises of two existing motorway sections, i.e. Elefsina – Korinthos (ELKO) 64 km long, and Patra by Pass (PbP) 18 km long, along with 120 km of the new Korinthos-Patra (KOPA) motorway, whose construction was completed in 2017, apart from the area of Rio I/C, that was completed in February 2018.

For the needs of the operation, the road has been divided into two Districts with facilities as shown in the following diagram (see Figures 11.1 and 11.2).

Figure 11.1. "Elefsina – Korinthos" Existing Section "Patra bypass" Existing Section.

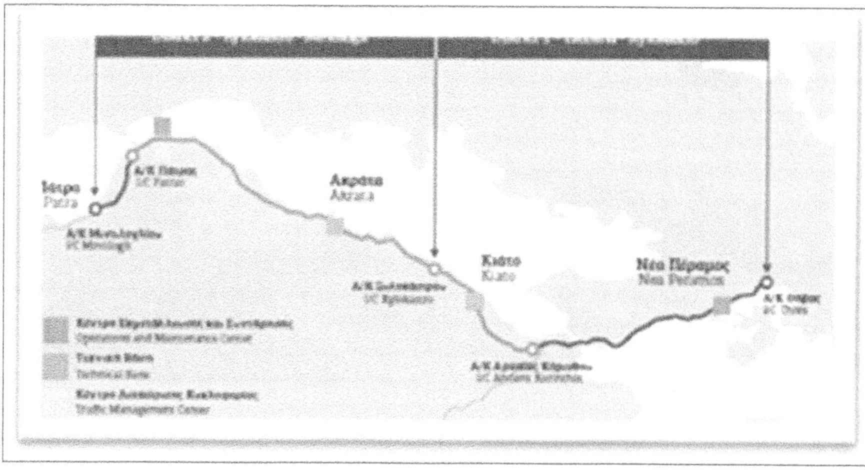

Figure 11.2. Project operation diagram into two districts.

The Pilot Site will be in the "Elefsina – Korinthos" section. The whole length of this section is 64 km with the following characteristics:

- Dual carriageway with 3 lanes (3,50 m width left lane, 3,75 m. with middle and right lane) & 1 Emergency Lane (varies from 2,50 to 4,50 m.) per direction, with concrete New Jersey safety barriers in the central axis of the motorway;
- Kakia Skala tunnel complex (5 tunnels of total length ~4,5 km) and 16 bridges;
- 2 large mainline toll plazas and 3 pairs of ramp toll plazas.

11.2.1.2 Spanish demo site: ACCIONA

The pilot project is set to launch on a section of the A2 Motorway managed by the company, specifically the R2–CM42 section starting from Madrid and ending at the border of the Guadalajara and Soria provinces in Spain. This operation also includes the traffic control center situated near Torija village. The roadway, governed by the Spanish National Road Authority, spans 77.5 km and is officially referred to as the "Public Works Contract for the Maintenance and Operation of the A-2 Dual Carriageway from Mile Marker 62.0 to 139.50. Section: R2 – L.P. Soria/Guadalajara."

This section features four lanes (two in each direction) and is situated in a region with a Continental-Mediterranean climate, known for its harsh winters and hot, dry summers. Due to high levels of heavy traffic, the road surface needs regular maintenance to remain in good condition. The A2, which connects Madrid to Barcelona, is a significant highway in Spain and is a component of the Trans-European Transport Network (TEN-T) and the CEF corridor.

Figure 11.3. Overview of the A2 motorway.

Figure 11.4. The Torija traffic control centre.

The contract is managed from the ACCIONA headquarters in Madrid and from the traffic control centre located in the small village of Torija. The traffic control centre is in charge of monitoring the motorway status, visualizing and assessing the data provided by CCTV, inductive loops, GPS-based fleets, weather stations, weigh-in-motion systems, etc. It is also the basecamp for all assets needed for maintenance (e.g., machinery).

11.2.1.3 French demo site: Transpolis

Transpolis is a proving ground of more than 80 ha which has been created by 5 entities among which Univ. Eiffel and has been opened officially in 2019. It is typically used to test autonomous vehicles in a secure and controlled environment thanks to several kilometers of road (notably 12 km in the "city area") and all reinforced concrete buildings. Many types of Vehicle to Everything (V2X) and Vehicle to Infrastructure (V2I) communication means are also available, as well as camera monitoring.

Concerning telecommunications, Transpolis is equipped with more than 320 km of optical fiber that allows access to an Ethernet network at almost any

Figure 11.5. Aerial View of the Transpolis site.

Figure 11.6. Several types of road markings at Transpolis.

point of the tracks. Thanks to an open LoRa network that covers the whole proving ground, a wide range of Internet of Things (IoT) sensors can be installed and used on Transpolis. Transpolis is also covered by a 5G network coming from an antenna located in the middle of the tracks. As far as energy is concerned, a private electrical distribution network guarantees power supply on the whole proving ground, especially in the "city area".

Transpolis will provide the proving ground for several use cases, namely:

- Detection and repair of road markings,
- Detection and repair of reinforced concrete cracks,
- Continuous V2I communication, between the vehicle and roadside units.

11.2.2 Demonstration Scenarios

In the HERON project, a series of demonstration scenarios are planned to illustrate the effectiveness of automated road maintenance and inspection procedures. The demonstrations are as follows:

1. Sealing Cracks: In this setup, UGVs equipped with sensors and the robotic arm are deployed to identify and seal cracks. They use high-resolution imagery to detect the cracks before sealing them with the necessary materials, helping to maintain a safer road surface.
2. Patching Potholes: This scenario demonstrates the project's ability to identify and repair potholes. The robotic units are equipped to detect potholes and fill them with appropriate materials, restoring the road's surface and preventing further decay.
3. Replacement of RUP Elements: This demonstration focuses on the removal and replacement of damaged or worn-out Removable Urban Pavement (RUP) elements. The robots are designed to identify these elements, remove them, and install new ones to maintain road safety standards.
4. Repainting of Road Markings: In this part, the HERON system showcases its capability to refresh faded or eroded road markings. The robots repaint the necessary lines and symbols with precision, ensuring clear and safe road demarcations.
5. Visual Inspections: This scenario involves the conducting of detailed visual inspections of road infrastructure. Using cameras and other sensors, they can collect comprehensive data on the condition of roads, aiding in informed decision-making regarding maintenance and repair tasks.

Across all these scenarios, an auxiliary function involves the deployment and removal of traffic cones to demarcate work zones, ensuring both the safety of the robotic units and the smooth flow of traffic. This aspect demonstrates the project's commitment to seamless integration into existing traffic systems while undertaking maintenance tasks.

Each demonstration is a testament to HERON's potential to improve road safety and lifespan through technological advancements and automation.

11.3 Computer Vision Toolkit

One of the main objectives of HERON is the identification of points of interest, i.e. the automatic detection and segmentation of cracks and potholes. For this task computer vision models have been developed based on the SegFormer [1] architecture. The SegFormer architecture, specifically designed for semantic segmentation tasks, merges the capabilities of transformers and convolutional neural networks (CNNs) to facilitate the categorization of different segments in an image. The structure begins with a hybrid backbone, a primary component responsible for extracting features from the input image. This backbone is constructed from a blend of convolutional and transformer layers, which are adept at extracting hierarchical features from images. Following this is the multi-level feature fusion, a vital element in aggregating features from diverse layers of the backbone. This helps in preserving both the high-level semantic information and the low-level spatial details, which are imperative for effective segmentation tasks.

At the heart of the SegFormer architecture lie the transformer layers, which play a pivotal role in modeling long-range dependencies between different regions of the

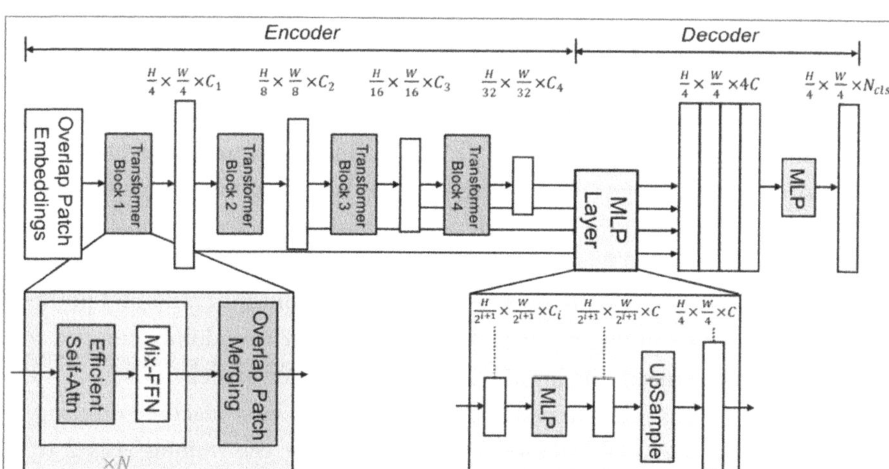

Figure 11.7. Segformer architecture.

Figure 11.8. An example of segmentation robustness on a corrupted (gaussian blur) sample from Cityscapes-C benchmark dataset. It is compared to a previously state-of-the-art model, DeepLabv3+. Source [1].

image. This facilitates an enhanced understanding of the semantic interrelationships between varying regions. Subsequently, the segmentation head comes into play, functioning to create the segmentation map. This segment assimilates the output from the transformer layers, transmuting them into segmentation masks. It might employ a straightforward convolutional layer or a complex structure to carve out detailed segmentation maps.

11.3.1 Performance Evaluation

The SegFormer architecture was trained using data captured by the project, as well as the "Cracks and Potholes in Road Images Dataset" by Passos *et al.* [7]. This is a publicly available dataset, and it was developed using images made available by the Brazilian National Department of Transport Infrastructure. It contains images of defects (cracks, potholes) in asphalted roads in Brazil, and it was made in order to be used for a study on the detection of cracks and potholes in asphalted roads. For the training process, the total of 2235 images have been split by a 80–20% ratio, for the training and validation data respectively from the cracks and potholes dataset. During the training process, the images has been randomly cropped to a 640 × 640 ratio to match the input size of the Segformer model.

The Segformer has been trained and evaluated in a Linux machine utilizing an NVIDIA1080Ti, of 12GB VRAM. The batch size for the training was 2 to 8, depending on the configurations of the various experiments run. The performance is illustrated in the figures below.

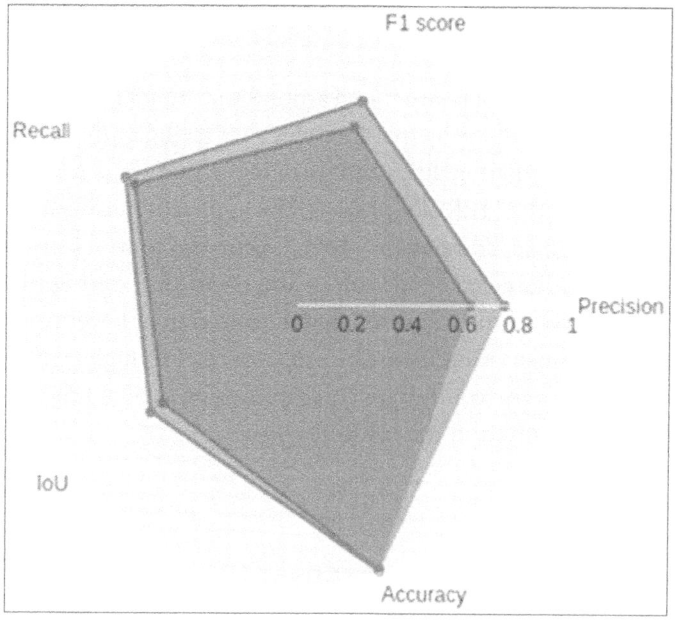

Figure 11.9. Visual representation of average metrics as presented in Table 4. Orange: Segformer, Blue: U-Net.

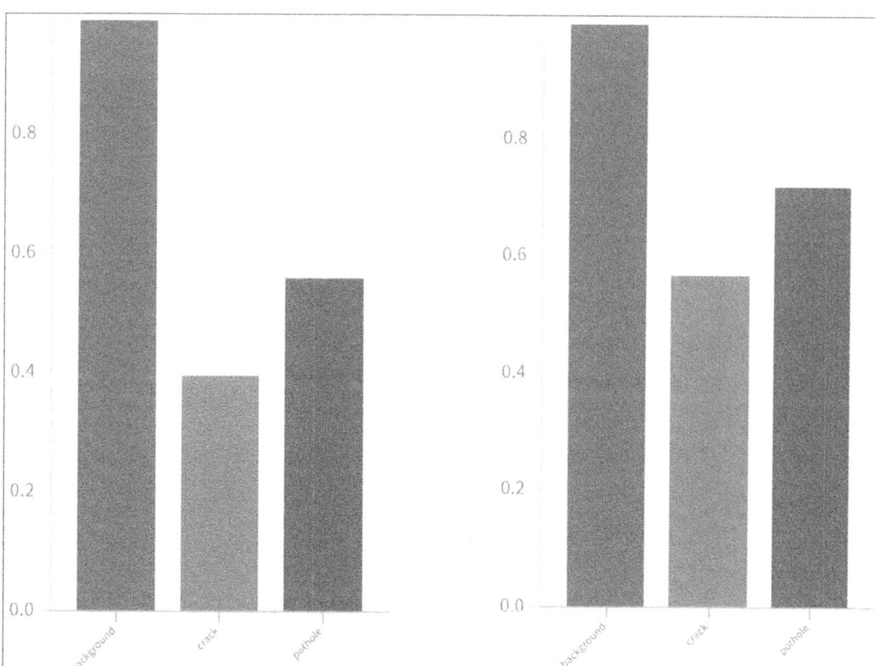

Figure 11.10. Segformer Performance for each individual class. Left: IoU score, right: F1 score.

11.4 Robotics Platform

The robotic platform system is structured around four fundamental elements. The control software houses the principal modules that facilitate the manipulation of the hardware system's functionalities and carry out actions imperative to fulfill the objectives outlined in the HERON project. The application software incorporates elements that amalgamate components derived from the control software segment, aiming to realize the targeted results of an intervention. The high-level planner serves to depict intervention tasks through the utilization of lower control software units, concurrently supervising them to ensure successful implementation. Lastly, the UGV interface acts as a conduit for communication between the UG and additional components involved in the HERON project.

11.4.1 UGV Interface

Considering the heterogeneous characteristics of the components within the HERON project and the projected creation of middleware proficient in interlinking them, it becomes evident that the robotic platform must facilitate communication with the entire system. Nonetheless, incorporating current software that

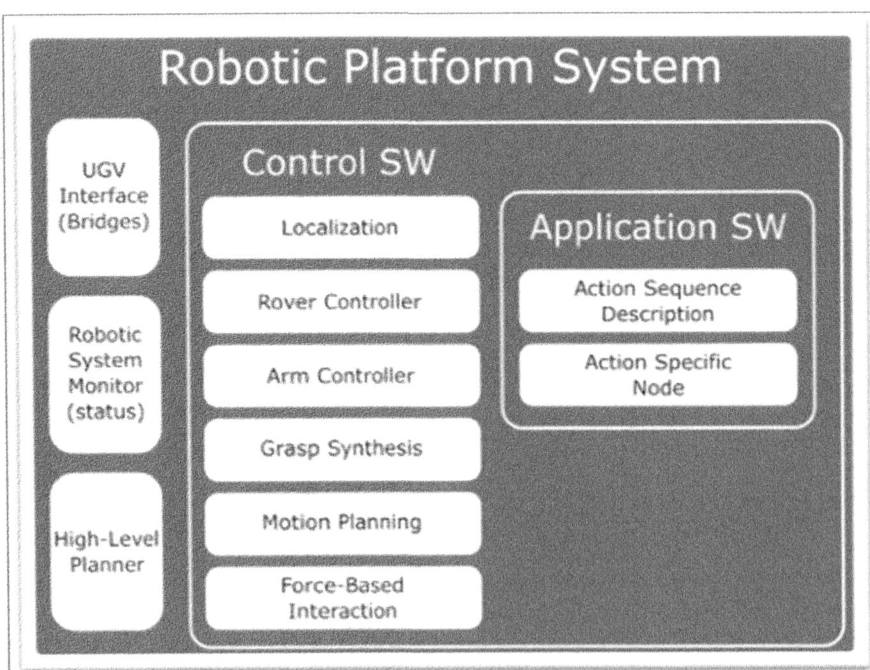

Figure 11.11. Overview of the robotic platform components.

encompasses certain foundational functionalities in the robot necessitates the retention of ROS-associated middleware at the robot's inferior levels. This component aims to unveil the robot's functionalities through various commands and signals, in conjunction with any necessary system feedback. Messages will be converted to ensure coherent communication between the specialized middleware and robot topics and services. Additionally, sensors pertaining to the robot will utilize this component to present the data available. Consequently, the UGV interface is poised to process an extensive variety of command calls and parameter setups, allowing it to engage with and adjust the system. This component's outcome consists of status feedback along with the data derived from the robot's sensors.

11.4.2 High-Level Planning

The high-level planning segment directs multiple actions accessible to the robot, orchestrating them to accomplish the intended task results. This procedure will leverage a domain-based planning representation to contemplate the interdependencies among actions and their pre and post-conditions, guaranteeing the creation of valid action sequences. The component functions based on a symbolic representation of both the environment and accessible actions. The target state is depicted using the identical symbolic representation, which a planning algorithm utilizes to delineate the series of actions needed to reach the goal.

11.4.3 Localization

The localization segment grants the robot the ability to gauge its location within a real-time evolving map of the environment. This system leverages cameras, inertial measurement units, and laser sensors to craft a volumetric metric depiction of the surroundings.

11.4.4 Robotic System Monitor

There exists a probability that the internal statuses of various sub-components need recognition and comprehension to convey the overarching status of the robotic platform. A comprehensive snapshot will also be presented, encompassing hardware availability and operational configurations, along with the requisite alerts and logs. In this vein, the component will assimilate input from both control software components and hardware specifics. This data is then employed to trigger alerts where needed while cataloging data for subsequent analysis and offering a summary of the system's condition.

11.4.5 Rover Controller

The HERON UGV is built upon a commercial Robotnik Robot, which will be customized for the maintenance tasks undertaken within the project. Fundamental movement functionalities form this component's nucleus, supporting various abstraction levels to accommodate diverse command types. In all instances, steering and traction motions will be directed to this component for conversion into motor actions. Commands will be accepted in the standard ROS movement command format, comprising linear velocity and steering angle directives, to be executed by an onboard closed-loop controller within the rover.

11.4.6 Arm Controller

The projection of incorporating a robotic arm to facilitate manipulation tasks within the project is anticipated. Several pre-existing arm options are under evaluation, contingent on the specific requirements of the use case. In all situations, proprietary drivers form the base layer for collaborative robotic arms, where manufacturers usually offer comprehensive APIs for robot interfacing. Yet again, a component translating user requirements into executable device instructions will be necessary. The arm controller will interpret inputs as either full motion plans, operation sequences, or individual directives. The outcome of any such input is the closed-loop adherence to the issued commands, accompanied by feedback.

11.4.7 Secondary Elements Controller

Supplementary operational elements that will be installed on the platform, like pumps, compressors, sprayers, raw material storage, level monitoring, and dispensing systems, will rely on a dedicated controller to synchronize actions with other HERON systems engaged in specific repair assignments, facilitating the mandated maintenance activities.

11.4.8 Grasp Synthesis

The grasp synthesis segment generates potential grasp points on object surfaces based on 3D reconstruction of the concerned object or scene. This technique demands a point cloud voxelized reconstruction of the object or scene in question, along with the robot's positioning and the manipulator's joint data. This module yields a list of positions and orientations for the end-effector on the object or scene's surface, promising a high grasp success probability.

11.4.9 Motion Planning

The motion planning segment devises trajectories that, when adhered to, guide the robotic system or its sections, such as the mobile base or manipulator, into a preferred setup. This planning mechanism necessitates an environmental representation in the form of an occupancy map. Besides the map, the mobile base and manipulator's locations are essential inputs to the planning structure. The objective is defined in terms of the state the robot aims to achieve with its mobile base and manipulator, along with any other pertinent data like operating volume, kinematic constraints, etc. The motion planner's output is a strategy, embodying a series of positions and speeds, executable by the lower-level control systems.

11.4.10 Force-Based Interaction

This component empowers the manipulator to engage with the environment in a way that deliberately fosters contact between the environment and the end-effector or manipulated object. The component demands the mobile base's position state and the manipulator's joint details. Along with this state information, the trajectory to be executed and any additional interaction constraints, such as maximum force, are required. The enactment of force-based interactions necessitates data concerning the forces and torques influencing the end-effector, and ideally the individual joints. This results in a high-frequency stream of control commands, consistently updated in alignment with the system's state alterations.

11.5 Visualisations & Improved Situational Awareness

The HERON system is designed to enhance situational awareness for principal parties involved in road maintenance (RM) and road inspection (RI) processes. Situational awareness involves recognizing the elements and events within the environment as they relate to time or space, understanding their significance, and forecasting their impending status. The strategy adopted by HERON is predicated on leveraging three primary components to offer users a continuous real-time (RT) data feed, thereby amplifying their situational awareness. These components are the Common Operational Picture (COP) module, an Augmented Reality (AR) application, and the Incidence Management & Decision Support System (IMS&DSS) application. The objective is to furnish decision-makers, operators, and field personnel with the comprehensive data necessary to streamline their operational plans and carry out effective road inspections and informed decision-making endeavors.

11.5.1 IMS & DSS

The IMS&DSS will be developed on the foundation of the lightweight client from the PANOPTIS H2020 project's IMS (refer to www.panoptis.eu). This foundation will be expanded to integrate with the HERON Middleware and to accommodate the unique specifications and business logic identified in the HERON case studies. Utilizing a containerized architecture, this system encapsulates software and its prerequisites in a discrete unit, facilitating streamlined server-side information processing and secure, efficient remote connections through a reverse proxy server. This server setup not only ensures load balancing and request tracking but also maintains a degree of anonymity for bolstered cybersecurity.

The system's infrastructure, housing a database server, adeptly manages diverse data inputs, processes it for uniformity and consolidation, and securely stores it in a distributed, resilient file system ready for retrieval and utilization by the IMS/DSS system.

Delving deeper, the IMS architecture comprises several elements:

1. Web Server: Dispatches the web application to browsers, administering the application business logic by coordinating user requests and actions with various servers and systems.
2. Reverse Proxy Server: Processes incoming http requests, enhancing them with HTTPS encryption and compression before rerouting them to the corresponding upstream server.
3. Database Server: Safeguards vital IMS information, including user data, access privileges, and sensor data acquired from middleware operations.
4. Video Management Server: Facilitates the ingestion and streaming of video content sourced from UxVs and road operator CCTV networks.
5. IMS/DSS App: A comprehensive web application fostering remote engagement with robotic resources and functionalities in field operations.

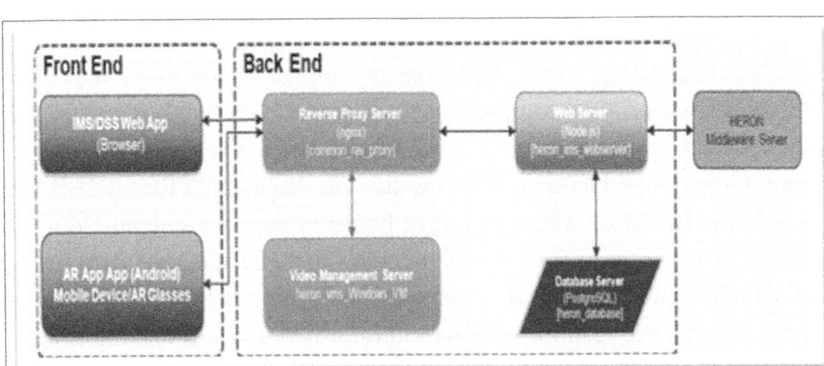

Figure 11.12. IMS system architecture.

6. AR App: A dynamic augmented reality application assisting users in pinpointing and analyzing structural road flaws or significant road components.

Together, these modules empower the IMS to cultivate and disseminate a Common Operational Picture (COP) among road inspection teams and pertinent road authorities, fostering collaborative efforts with various regional and local stakeholders as necessary. To sustain coherent communication and streamline processes, the IMS will employ specific protocols to nurture multi-level and multi-actor interactions, ensuring a unified approach to task execution

11.5.2 COP

As highlighted earlier, the HERON initiative enhances situational awareness capabilities by distributing detailed information obtained from all HERON sensors. This information is complemented by data from fusion processes, navigation modules, and positioning and advanced planning units, all of which are encompassed in what is termed the Common Operational Picture (COP). This centralized virtual representation of the HERON Robotic platform controller acts as a vital resource for robot operators and decision-makers in road inspection companies, enabling them to organize their assignments proficiently.

The COP's complex elements will be segmented into various layers and categories of information, creating a flexible system that operates on a "need-to-know" basis. This is in anticipation of the diverse user profiles and roles that will interact with the HERON tools and services through the AR, IMS, and DSS systems.

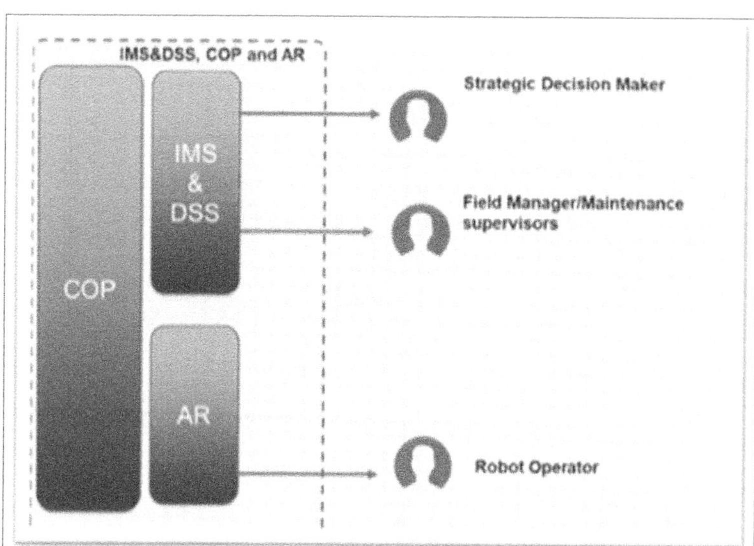

Figure 11.13. IMS, COP, AR user and roles.

11.5.3 AR App

The AR system will undertake the task of delivering real-time graphical data on the environment surrounding the robot operators. The AR application, specifically designed for Android-based AR devices, will facilitate the visualization of existing defects, automating the detection and severity classification of pavement flaws. Furthermore, the Android AR application will employ 3D model overlays to reveal potentially concealed structural elements that might influence the maintenance procedures or cause further damage. The display of functional elements will also be accessible through relevant commands.

11.5.4 Middleware and Data Fusion

The overall middleware and DF architecture of HERON are presented in the following figure.

The HERON middleware is structured to integrate data from various HERON elements and sensors, interfacing efficiently with the application layer while maintaining high reliability and scalability. This structure is divided into two main layers.

The first layer focuses on essential data pre-processing (Figure 11.14), consolidating information from multiple sources into unified data models using

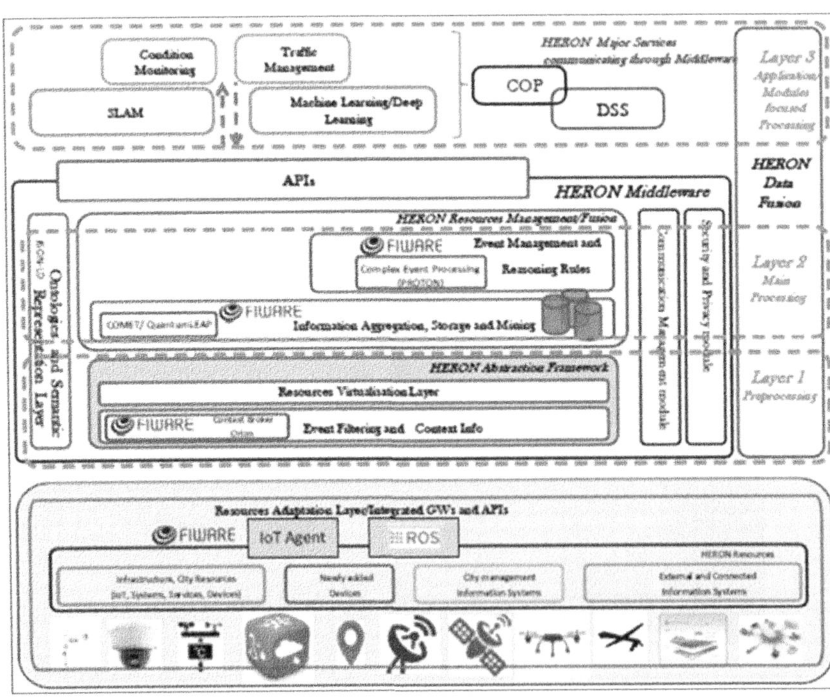

Figure 11.14. Schematic diagram of HERON middleware.

established FIWARE data models and JSON schemas for a straightforward key-value representation of context data.

The second layer is concerned with data storage, processing, and forwarding it to the application layer. Here, data undergoes virtualization into objects and is normalized and stored. This layer manages the primary processing of data from Layer 1, where it works closely with Resource management to create additional events and manage the data fusion process. Through Event Management, it handles, categorizes, and processes events and data from different sources, making them readily accessible via the API to the DSS system and other modules that need extra data. It also makes raw data available for further processing by high-level modules and applications.

To ensure the safety and integrity of the data, the middleware only accepts and stores data from trusted sources, allowing access to authorized requests only. It employs a policy-based management framework with mechanisms for data encryption, access control, and privacy protection, supplemented with tools for intrusion detection and prevention.

Finally, the system will establish necessary interfaces and protocols to communicate with different data sources and user services, managing aspects such as time synchronization, scheduling, communication path selection, fault tolerance, and traffic shaping seamlessly.

11.6 Conclusions

In conclusion, the HERON project stands as a beacon of innovation and advancement in the realm of road infrastructure maintenance, proposing a holistic, automated system buttressed by an autonomous ground robotic vehicle. This groundbreaking initiative not only pioneers the use of cutting-edge technology in infrastructure upkeep but also serves as a catalyst for enhanced safety, efficiency, and adaptability across diverse transportation networks.

Central to the HERON project is its innovative approach to road infrastructure maintenance. The integration of an autonomous ground robotic vehicle signifies a paradigm shift, fostering a more proactive and responsive system capable of adapting to the myriad demands of contemporary transport infrastructures. This transformative vehicle is envisioned to be the nucleus of the HERON system, orchestrating a seamless and continuous vehicle for infrastructure data exchange. By doing so, it aims to significantly heighten user safety through the constant monitoring and rapid response to changing road conditions.

Moreover, the HERON initiative is designed to fully harness and maximize the capabilities of modern technology to foster adaptability across various transport infrastructures. This adaptability is seen as a cornerstone in reducing the incidence of fatal accidents, a pressing issue that plagues road networks globally. Furthermore, by introducing a streamlined, automated approach, the project stands to significantly mitigate maintenance costs and traffic disruptions, thus fostering a more fluid and efficient transport environment.

Looking towards the future, the HERON project is poised for further evolution and refinement. The forthcoming phases anticipate the meticulous integration of individual components such as sensors, actuators, and tools, each playing a critical role in the system's overall functionality and efficiency. These components, working in synergy, are expected to bring unprecedented levels of precision and reliability to infrastructure maintenance.

Furthermore, the project embarks on the journey towards the on-site demonstration of the HERON system, a critical milestone that stands to validate the potential and efficacy of this revolutionary approach. This demonstration aims to showcase the real-world applicability and benefits of the HERON system, potentially serving as a blueprint for similar initiatives globally.

In essence, the HERON project represents a promising stride towards a safer, more efficient, and responsive road infrastructure network, marrying technology and innovation to meet the complex demands of modern societies. It serves as a testament to the potential of collaborative efforts in bringing about substantial advancements in road infrastructure maintenance, potentially ushering in a new era of safety and efficiency on the roads that connect us all.

Acknowledgments

This work has received funding from the European Union's Horizon 2020 Research and Innovation Programme under grant agreement No 955356 (Improved Robotic Platform to perform Maintenance and Upgrading Roadworks: The HERON Approach).

References

[1] Katsamenis, I., Bimpas, M., Protopapadakis, E., Zafeiropoulos, C., Kalogeras, D., Doulamis, A., … and Lopez, R. (2022, June). Robotic maintenance of road infrastructures: The heron project. In Proceedings of the 15th International Conference on PErvasive Technologies Related to Assistive Environments (pp. 628–635).

[2] InfraROB. Maintaining integrity, performance and safety of the road infrastructure through autonomous robotized solutions and modularization. https://infrarobproject.com/, 2021. [Online; accessed 01-September-2023].

[3] González-Collazo, S. M., del Río-Barral, P., Balado, J., and González, E. (2022). Detection of direct sun glare on drivers from point clouds. Remote Sensing, 14(6), 1456.

[4] OMICRON. Towards a more automated and optimised maintenance, renewal and upgrade of roads by means of robotised technologies and intelligent decision support tools. https://omicronproject.eu/, 2021. [Online; accessed 01-September-2023].

[5] Sevilla, I., Chrobocinski, P., Barmpas, F., Schmidt, F., Kerle, N., Kostaridis, A., ... and Russotto, R. (2018, November). Improving Resilience of Transport Instrastructure to Climate Change and other natural and Manmande events based on the combined use of Terrestrial and Airbone Sensors and Advanced Modelling Tools. In CONAMA2018 (pp. 11-p).

[6] SegFormer: Simple and Efficient Design for Semantic Segmentation with Transformers, demonstration link: https://www.youtube.com/watch?v=J0MoRQzZe8U.

[7] Passos, B.T., Cassaniga, M., Fernandes, A.D.R., Medeiros, K.B. and Comunello, E., 2020. Cracks and potholes in road images. Mendeley data, 4, p. 2020.

Epilogue

In an era when civil infrastructures are strongly ageing, digital technologies show a vast increase in a large number of applications and industrial domains. Being aware of recent approaches to cope with civil structures ageing is a first step towards a more effective maintenance and inspection process and plan that can maintain the original integrity of the structures. Without this imperative step, there are high risks including financial, health, safety, performance and other with high probability and impact that can in turn affect the related ecosystems.

This book summarizes recent technologies that can be used alone or in combinations to provide a more robust and resilient infrastructures' system in EU and worldwide. The work presented includes mostly results from research projects related to industrial inspection and maintenance of civil structures but includes also results with high commercial prospects and direction. More specifically, the following concepts, technologies and approaches are included in the book chapters:

- Integrated Inspection and Maintenance (I&M) Platform for Critical Infrastructures as an outcome of the PILOTING research project, being a software platform that performs orchestration of the different modules developed in the project. The I&M platform is a software application including several operational layers of the PILOTING system completing the backbone infrastructure.
- AI algorithms for corrosion detection in oil and gas refineries as another outcome of the PILOTING project including computer vision algorithms for corrosion and other defects detection in pipes and tanks of the oil and gas industry.
- Precise, laser scanning based approach for tunnel shape capturing and 3D deformation detection and assessment, also developed in the PILOTING project. This includes the 3D laser scanner setup, configuration, approach as

well as models comparison approaches to compare older and current tunnel internal structural images.

- Transforming Engineering Inspections through Advanced UAVs as modern approaches performed in the PILOTING project, including several robotics platforms. Such approaches include results and achievements of the PILOTING project with Unmanned Aerial Vehicles (UAVs) developed to execute inspections in challenging and inaccessible environments.
- A visualization and intelligence portal including an end-to-end workflow of digital inspections in challenging environments including an intuitive user interface also developed in the PILOTING project.
- Autonomous inspection and maintenance operations as also part of inspection and maintenance activities with robotics, having the potential to reduce the cost, increase the quality and enable new functionalities. Robust and accurate navigation algorithms for ground vehicles are include as a key enabler for safe operations and reliable location tagging of inspection data.
- Other critical infrastructures inspection activities are also included, summarizing activities of Application of Intelligent Aerial Robots to the Inspection and Maintenance of Electrical Power Lines.
- Other applications include the promotion of the use of Robotics in the Inspection and Maintenance of Offshore Wind as another critical infrastructure paradigm of major importance with significant challenges in manual inspection approaches.
- Application domains also extend to Intelligent Road Management Platforms aimed at enhancing the construction, maintenance, renovation, and rehabilitation of the European road network as well as integrated automated systems to perform maintenance and upgrading roadworks, such as sealing cracks, patching potholes, asphalt rejuvenation, autonomous replacement of CUD elements and painting markings, but also supporting the pre/post-intervention phase including visual inspections and dispensing and removing traffic cones in an automated and controlled manner.

The above highly innovative techniques are already at a stage of TRL6-7 (or will be at the end of the respective efforts) as all of them have been demonstrated and validated in various industrial scenarios and cases. Following recent research and industrial results especially including recent digital transformation solutions involving IoT, digital twins, BIM or other related technologies, is considered of primal importance and presents high value in civil inspection and maintenance.

Index

3D data visualization, 128, 132–136, 138, 151, 156
3D digital twin, 128, 138, 162
3D model, 255
3D modelling, 88, 89, 98, 99
aerial co-workers, 184, 199
aerial dual arm manipulator, 192, 193
aerial inspection, 244
aerial manipulation, 184, 185, 189, 192, 201
aerial manipulators, 183, 184, 189, 197
aerial robotics, 184
aerial robots, 104, 119, 120, 122, 124, 183, 184, 187
ageing infrastructures, 88
AI-assisted annotation, 62
annotation, 62, 70, 72, 76, 77, 79, 83
API, 22, 24–26, 28–35, 39, 40, 42–44, 46, 49–51
application of artificial intelligence, 184
artificial intelligence, 62, 69
asset inspection, 17–19, 21, 23–29, 33, 34, 36–40, 43, 45, 46, 48, 53, 55–57
asset inspection and maintenance, 156
asset management software, 132, 134, 137, 139
automated, 61, 62
autonomous flight, 116
autonomous vehicles, 165, 166, 169, 172, 174, 180
autonomy, 211
BPMN, 261, 262, 264, 268
building information modelling, 99
computer vision, 62, 65, 67
construction sites, 62, 64
contact inspection, 104, 105, 107, 108
convolutional neural networks, 61, 62, 69
corrosion, 62, 64, 65, 69, 76, 81
critical infrastructures, 17, 19, 56
CVAT (computer vision annotation tool), 62
CycleGans, 61, 63, 64, 70, 72, 73, 75, 77
data analysis, 133, 136, 146, 147, 151, 158
data augmentations, 66, 70, 82
data harmonization, 17, 19, 34, 36, 38, 40, 42, 44–46, 48, 52
data management system, 17, 19, 22, 26, 34, 39, 40, 42, 43, 45, 46, 57
data model, 21–25, 34, 40, 51–54
data visualization tools, 136, 138, 143, 145, 147
data-driven decision making, 130

decision making, 250
decision support system, 249, 267, 268
decision support tool, 247, 249, 250, 258, 259, 264, 266, 268
deep learning, 61–64, 67, 72
deformation, 87–89, 91, 93, 96–99
detection, 61, 62, 64, 65, 69, 76, 78, 80, 81, 85
digital inspection solution, 134, 141, 163
digital transformation, 128, 133
digital twin, 247–250, 258–262, 264–269
digitalization, 128, 130, 267
drones, 104
electrical power lines, 183–185
end-to-end portal, 127, 162
formation control, 199, 201
GNSS-free navigation, 106
human-machine interfaces, 184, 185, 197, 201
I&M, 90
image annotation, 151, 162
IMR, 209–213, 220–223, 231, 234, 235, 237, 238, 242, 243
infrastructure, 61, 62
inspection, 61, 62, 87, 88, 90–96, 99, 165, 166, 169–172, 174, 177–180, 183–188, 195, 197, 199, 201, 202, 247–258, 260, 263, 265–268
inspection and maintenance, 17–19, 56, 57
inspection at height, 104, 105
inspection data management, 129, 130, 138, 146, 148, 151
inspection planning, 128, 130, 146
inspection report, 136, 148, 153, 154, 158–160, 162
inspection workflow, 127, 128, 132
installation of bird diverters, 190, 193
installation of cable spacers, 190
intelligent asset monitoring, 137, 155
intelligent platform, 248, 249, 268
IoT, 128, 136, 155, 156, 265, 266
IoT (Internet of Things), 25, 29, 35, 48, 50
Keycloak, 30–33
laser scanning, 93, 94, 96, 99
lidar, 168, 169, 174
line detection, 186, 187
line tracking, 185, 186
localization, 165–170, 174, 179
long-range mapping, 189
magnetic battery chargers, 183, 185, 195, 196, 201
maintenance, 165, 166, 172, 177, 184, 185, 189, 195, 197, 199, 201, 247–251, 258, 260, 261, 263–268
manipulation, 210, 238, 241, 243
mapping, 166, 168–172, 174–177
maritime robotics, 209, 213, 214, 220, 242
MaskRCNN, 62, 66, 69, 78–82
MAVLink, 56
monitoring, 251–253, 257, 260, 265, 266, 268
MQTT, 29, 48
multi-domain inspection, 245
multi-UAV, 250, 253, 256–258, 268
navigation, 165–167, 169–172, 174, 175, 179, 180
NGINX, 35, 40
O&M, 207–211, 213, 214, 220, 222, 236, 237, 242, 243
offshore wind, 207–210, 213–216, 219, 220, 222, 223, 226, 234, 235, 238, 242, 243
OGC, 24
ontology, 260–263, 268
open access license, 22, 34, 35, 40, 51, 56
operation optimization, 207, 220, 222, 236, 243
pavement state, 265, 268
perception, 210, 223–226, 233, 239, 243

planning, 247, 250, 252, 256, 258, 264, 265, 268
point cloud, 133, 134, 138–140, 266
power line grids, 183–186, 190, 192, 193, 201
precise localization, 116
predictive maintenance, 237, 243
RabbitMQ, 35, 42, 46
RCS (Robot Control Station), 35, 39–41, 45, 51, 56
reality capture, 247, 250
refineries, 18, 48, 104, 105, 110
refinery, 165, 169, 170, 172, 174
road tunnels, 18, 36
robotic systems, 127, 135, 145, 146, 162
RPAS, 105, 111
safety monitoring with aerial robots, 183–185, 199, 201
scanning, 87, 93, 94, 96, 99
sensor and target installation, 104–110, 112, 114–120, 123
sensor fusion, 168
SensorThings, 22, 24–26, 29–33, 35, 40, 42, 48–50
SLAM, 165, 166, 168–170, 174, 175, 177, 178, 180
structural assessment, 87, 91
structural health monitoring, 265, 268
structural integrity, 62
supervised object detection, 62
supervisory control centre, 214, 217, 234, 243
swin transformers, 62, 67, 68, 80
technology certification, 207, 220
technology validation, 220
test centre, 208, 209, 213–217, 219–222, 224, 226, 228, 230–232, 234, 237, 239, 243
tunnel, 87–99, 165, 166, 168, 169, 174–179
UAV, 247–254, 257–259, 263, 266, 268, 269
UAV battery charging from power lines, 184
UAVs, 103, 104
UAVs management tool, 252
UGV, 169, 170, 172, 174, 175, 177, 179
UML, 261–263, 268
underwater inspection, 227, 238
unmanned aerial vehicles, 184
UT inspection, 107, 110, 111
valve manipulation, 172, 173
viaducts, 18, 36, 48, 104, 105, 111, 113, 116, 118
visual inspection, 104, 105, 111, 113–116, 119, 124
wireless sensor network, 29
YOLO, 62, 65, 67, 76, 77, 82

About the Editor

Konstantinos Loupos

In the capacity of editor for this publication, Mr. Konstantinos Loupos, whose remarkable expertise and substantial contributions to the fields of microelectronics, embedded systems, IoT/ICT technologies, and Cybersecurity have created a notable presence in the realm of technological innovation and research.

Mr. Loupos holds a distinguished academic record, encompassing an MBA from the Hellenic Open University (Greece), an M.Sc. in Microelectronics Systems Design with distinction from the University of Southampton (United Kingdom), and an M.Eng. in Electronic and Electrical Engineering from the University of Manchester (United Kingdom). His educational accomplishments serve as a testament to his commitment to academic excellence and his relentless pursuit of knowledge.

With a multifaceted career spanning in various domains, Mr. Loupos has accumulated extensive practical experience and knowledge in an array of critical areas. His research interests and activities encompass IoT/ICT systems, Cybersecurity, Sensors and Systems for Structural Health Monitoring (including tunnels, structures, bridges, and more), Robotics for Civil Infrastructure Inspection, Security Systems (comprising sensors and communications), Water Demand Management, and Wireless Sensor Networks. Additionally, his expertise extends to Application Specific Embedded Systems, further showcasing his versatility in the technological landscape.

Mr. Loupos's exceptional dedication to research is evident through his active participation in over 35 EC co-funded research projects, spanning from FP5 to H2020, Horizon Europe and beyond. His contributions have spanned diverse sectors, including transport, security, health, climate, cultural heritage, ICT, and nano-materials, highlighting his commitment to advancing knowledge and fostering innovation. Throughout his career, he has held significant roles, ranging from

Project Coordinator to Technical Manager and Leader of Development teams, all of which underscore his leadership and management capabilities.

Furthermore, Mr. Loupos's scholarly endeavors are reflected in his impressive portfolio of over 70 publications in conference proceedings, journals, and book chapters. His dedication to the academic community extends beyond publications, as he has actively served as an organizer, technical chair, and member of organizing committees for numerous scientific conferences, such as IoT Week, Global IoT Summit, International Conference on Availability, Reliability and Security, International Physical Internet Conference, Transport Research Arena, IoT World, IoT Solutions World Congress, IEEE World Forum on IoT, Globecomm, IoTi4, Mobisec and more.

In addition to his research and academic contributions, Mr. Loupos has played a crucial role as a formal evaluator (expert) of EC projects in various programs, including FP7, H2020, HEU, EUKERA/EUROSTARS, MCST, MARTERA, COST, ERANET, and MED. Simultaneously, he has undertaken the role of project reviewer (external technical expert) for ongoing Horizon 2020 projects, thereby contributing to the assessment and advancement of cutting-edge research initiatives.

Mr. Loupos's commitment to professional development is evident in his certifications as a Project Management Professional (PMP), Scrum Expert, and GDPR expert (Certified DPO). His active memberships in organizations such as the EUROPOL EC3 cyber security group, European Cybercrime Center, Secure Platform for Accredited Cybercrime Experts (SPACE), EUROPOL Data Protection Experts Network (EDEN), the Alliance for Internet of Things Innovation (AIOTI), and SPRINT Robotics further reflect his engagement in fostering collaboration and addressing contemporary challenges in the technological landscape.

Currently, Mr. Loupos serves as the R&D Director at INLECOM, where he holds positions on the board of directors and as the R&D strategy keeper and business development. His involvement in numerous research projects, both as a coordinator and technical manager, continues to shape and advance technological innovation in diverse domains.

In alignment with the theme of this book, Mr. Loupos possesses extensive experience in structural health monitoring initiatives, encompassing sensors, wireless sensor networks, civil structures monitoring, and 3D deformation assessment. His pivotal roles in projects dating back to 2010, including MONICO, MEMSCON, RECONASS, ROBO-SPECT, SENSKIN, INACHUS, INTERQ, PILOTING, and other projects, underscore his profound impact on the field of structural health assessment and monitoring.

Contributing Authors

Luca Aceto
RINA, Palazzo R, Strada 7, Via Gran S. Bernardo, 20089 Rozzano
MILANO – ITALY
luca.aceto@rina.org

Angelos Amditis
Institute of Communication and Computer Systems, National Technical University of Athens
157 73 Athens
a.amditis@iccs.gr

Olov Andersson
ETH Zurich, Switzerland
olov.andersson@mavt.ethz.ch

Arnaud Andrianavalomahefa
Quasset, Seinstraat 4, 1223 DA Hilversum
arnaud.andrianavalomahefa@quasset.com

Paula López Arévalo
CEMOSA, Street Benaque 9, 29004, Málaga, Spain
paula.lopez@cemosa.es

Elena Avatangelou
Innov-acts Ltd, Cyprus
elena.avatangelou@innov-acts.com

Nikolaos Bakalos
Institute of Communication and Computer Systems,
Greece
bakalosnik@mail.ntua.gr

Aníbal Ollero Baturone
Robotics, Vision and Control Group, University of Seville, Escuela Superior de Ingenieros, Camino de los Descubrimientos s/n, 41092 Seville (Spain)
aollero@us.es

Maruša Benkic
CORTE, Belgium
m.benkic@corte.be

Dimitrios Bilionis
RISA Sicherheitsanalysen GmbH, Germany
d.bilionis@risa.de

Stjpan Bogdan
LARICS. Department of Control and Computer Engineering. University of Zagreb, Unska 3,

Zagreb 10000 (Croatia)
Stjepan.Bogdan@fer.hr

Cesar Cadena
ETH Zürich, Leonhardstrasse 21, Zurich, Switzerland
cesar.cadena@ethz-asl.ch

Daniel Filipe Campos
Centre for Robotics and Automation Systems, INESC TEC, R. Dr. Roberto Frias, 4200-465, Porto, Portugal
daniel.f.campos@inesctec.pt

Miquel Cantero
Robotnik Automation SLL, Spain
mcantero@robotnik.es

Elena Chiggiato
Quasset, Seinstraat 4, 1223 DA Hilversum
elena.chiggiato@quasset.com

Giovanni Cioffi
University of Zurich, Andreasstrasse 15, 8050 Zurich (Switzerland)
cioffi@ifi.uzh.ch

Rafael Claro
Centre for Robotics and Automation Systems, INESC TEC, R. Dr. Roberto Frias, 4200-465, Porto, Portugal
rafael.marques@inesctec.pt

Marianna Corinaldesi
Quasset, Seinstraat 4, 1223 DA Hilversum
marianna.corinaldesi@quasset.com

Alejandro Murcia Crespo
CATEC (Advanced Center for Aerospace Technologies), C/Wilbur y Orville Wright n19, 41309, La Rinconada (Seville), Spain
amurcia@catec.aero

José Ramiro Martínez de Dios
Robotics, Vision and Control Group, University of Seville, Escuela Superior de Ingenieros, Camino de los Descubrimientos s/n, 41092 Seville (Spain)
jdedios@us.es

Emad Ebeid
SDU Digital and High-Frequency Electronics, University of Southern Denmark, Campusvej 55, Odense M, (Denmark)
esme@sdu.d

Ioanna Fergadiotou
Inlecom Innovation, 145 61 Kifissia, Greece
Ioanna.fergadiotou @inlecomsystems.com

João Formiga
EDP New, Rua Particular à R. Cidade de Goa 2, 2685-039 Sacavém, Portugal
joao.formiga@edp.com

Nikos Frangakis
iKnowHow SA, Greece
nfrangakis@iknowhow.com

Martin Eek Gerhardsen
SINTEF Digital, Forskningsveien 1, Oslo, Norway
Martin.Eek.Gerhardsen@sintef

Ilias Gkotsis
Satways Ltd, Greece
i.gkotsis@satways.net

Rafael Caballero González
CATEC (Advanced Center for Aerospace Technologies), C/Wilbur y Orville Wright n19, 41309, La Rinconada (Seville), Spain
rcaballero@catec.aero

Shashank Govindaraj
Space Applications Services, Leuvensesteenweg 325, 1932 Sint-Stevens-Woluwe (Brussels Area), Belgium
shashank.govindaraj@spaceapplications.com

Francisco Javier Pérez Grau
CATEC (Advanced Center for Aerospace Technologies), C/Wilbur y Orville Wright n19, 41309, La Rinconada (Seville), Spain
fjperez@catec.aero;

Marco Antonio Montes Grova
CATEC (Advanced Center for Aerospace Technologies) C/Wilbur y Orville Wright n19, 41309, La Rinconada (Seville), Spain
mmontes@catec.aero

Yannis Handanos
Olympia Odos Operation SA, Greece
yhandanos@olympiaoperation.gr

José Solís Hernández
CEMOSA, Street Benaque 9, 29004, Málaga, Spain
jose.solis@cemosa.es

Viet Duong Hoang
SDU Digital and High-Frequency Electronics, University of Southern Denmark, Campusvej 55, Odense M, Denmark
vdh@sdu.dk

Peter Ørnulf Ivarsen
SINTEF Digital, Forskningsveien 1, Oslo, Norway
peter.ivarsen@sintef.no

Aikaterini Karagiannopoulou
Institute of Communication and Computer Systems, National Technical University of Athens 157 73 Athens
katerina.karagiannopoulou@iccs.gr

Athanasios Karydis
INLECOM INNOVATION, Tatoiou 11, 14561, Kifisia, Greece
thanos.karydis@inlecomsystems.com

Iason Katsamenis
Institute of Communication and Computer Systems, Greece
iasonkatsamenis@mail.ntua.gr

Julian Keller
ETH Zürich, Leonhardstrasse 21, Zurich, Switzerland
julian.keller@mavt.ethz.ch

Valia Konstantopoulou
INLECOM INNOVATION, Tatoiou 11, 14561, Kifisia, Greece
valia.konstantopoulou@inlecomsystems.com

Gerasimos Kouloumbis
INLECOM INNOVATION, Tatoiou 11, 14561, Kifisia, Greece
gerasimos.kouloumbis@inlecomsystems.com

Vincenzo Lippiello
CREATE Consortium and University of Naples Federico II, Via Claudio 21, 80125, Napoli (Italy)
vincenzo.lippiello@unina.it

Konstantinos Loupos
INLECOM INNOVATION, Tatoiou 11, 14561, Kifisia, Greece
Konstantinos.loupos@inlecomsystems.com

Henrik Lundkvist
SINTEF Digital, Klæbuveien 153, Trondheim, Norway
Henrik.lundkvist@sintef.no

Michele Marolla
CREATE Consortium and University of Naples Federico II, Via Claudio 21, 80125, Napoli (Italy)
mic.marolla@gmail.com

Juan Manuel Marredo
Advanced Center for Aerospace Technologies (CATEC), Wilbur and Orville Wright 19 - 41309 La Rinconada, Sevilla (Spain)
jmmarredo@catec.aero

João V Amorim Marques
Centre for Robotics and Automation Systems, INESC TEC, R. Dr. Roberto Frias, 4200-465, Porto, Portugal
joao.marques@inesctec.pt

Aníbal Matos
Centre for Robotics and Automation Systems, INESC TEC, R. Dr. Roberto Frias, 4200-465, Porto, Portugal
andry.m.pinto@inesctec.pt

Eeva Mikola
VTT, P.O. Box 1000, FI-02044 VTT Finlandeeva.mikola@vtt.fi

Mohammed El Mobachi
Principle Power, 1330 rue Jean-René Guilibert Gauthier de la Lauzière, 13290 Aix en Provence, France
mmobachi@principlepowerinc.com

Carlos Martín-Portugués Montoliu
Acciona Construcción SA, Spain
carlos.martinportugues.montoliu@acciona.com

Richard J. D. Moore
SINTEF Digital, Forskningsveien 1, Oslo, Norway
Richard.Moore@sintef.no

Helen Oleynikova
ETH Zurich, Switzerland
helen.oleynikova@ethz-asl.ch

Anibal Ollero
GRVC Robotics Lab. Universidad de Sevilla, ETS Ingeniería, Camino Descubrimientos, 41012 Sevilla (Spain)
aollero@us.es

Maria Inês Pereira
Centre for Robotics and Automation Systems, INESC TEC, R. Dr. Roberto Frias, 4200-465, Porto, Portugal
Maria.i.pereira@inesctec.pt

Robert Pìnièka
Czech Technical University in Prague, Karlovo námìstí 293/13, 121 35 Praha (Czech Republic
penicrob@fel.cvut.cz

Andry Maykol Pinto
Centre for Robotics and Automation Systems, INESC TEC, R. Dr. Roberto Frias, 4200-465, Porto, Portugal
andry.m.pinto@inesctec.pt

Jonathan Prevosto
Exail, 220 Rue Georges Onslow,

34070 Montpellier, France
jonathan.prevosto@exail.com

Eftychios Protopapadakis
University of Macedonia, Greece
eftprot@uom.edu.gr

Safidy Ratsimbazafy
Quasset, Seinstraat 4, 1223 DA Hilversum
safidy.ratsimbazafy@quasset.com

Noemi Jiménez Redondo
CEMOSA, Street Benaque 9, 29004, Málaga, Spain
noemi.jimenez@cemosa.es

Balázs Némedi Régens
Street Prielle Kornelia 19/D, 1117, Budapest, Hungary
balazs.nemedi@regens.com

Victor Reijgwart
ETH Zürich, Leonhardstrasse 21, Zurich, Switzerland
victor.reijgwart@mavt.ethz

David Ribas
IQUA Robotics, Pic de Peguera 11, 17003 – Girona, Spain
david.ribas@iquarobotics.com

Pere Ridao
Univeresitat de Girona, Parc Científic i Tecnològic de la Universitat de Girona, 17003 – Girona
pere.ridao@udg.edu

Konstantinos Routsis
Institute of Communication and Computer Systems, National Technical University of Athens 157 73 Athens
konstantinos.routsis@iccs.gr

Fabio Ruggiero
CREATE Consortium and University of Naples Federico II, Via Claudio 21, 80125, Napoli (Italy)
Stjepan.Bogdan@fer.hr

Andreas El Saer
INLECOM INNOVATION, Tatoiou 11, 14561, Kifisia, Greece
andreas.elsaer@inlecomsystems.com

George Sarigiannis
INLECOM INNOVATION, Tatoiou 11, 14561, Kifisia, Greece
george.sarigiannis@inlecomsystems.com

Martin Saska
Czech Technical University in Prague, Karlovo námìstí 293/13, 121 35 Praha (Czech Republic)
Martin.Saska@cvut.cz

Georgios Sarigiannis
Inlecom Innovation, 145 61 Kifissia, Greece
george.sarigiannis@inlecomsystems.com

Davide Scaramuzza
University of Zurich, Andreasstrasse 15, 8050 Zurich (Switzerland)
sdavide@ifi.uzh.ch

Franziska Schmidt
Université Gustave Eiffel, France
franziska.schmidt@univ-eiffel.fr

Ambar Garofano Soldado
Robotics, Vision and Control Group, University of Seville, Escuela Superior de Ingenieros, Camino de los Descubrimientos s/n, 41092 Seville (Spain)
agarofano@us.es

Miguel Ángel Trujillo Soto
CATEC (Advanced Center for Aerospace Technologies), C/Wilbur y Orville Wright n19, 41309, La Rinconada (Seville), Spain
matrujillo@catec.aero

Jan Sramota
SINTEF Digital, Klæbuveien 153, Trondheim, Norway
Jan.Sramota@sintef.no

Luc Stakenborg
Quasset, Seinstraat 4, 1223 DA Hilversum
luc.stakenborg@quasset.com

Jaap-Jan Stoker
ABB, Frankeneng 15, NL – 6716 AA Ede, Netherlands
jaap-jan.stoker@nl.abb.com

Alejandro Suarez
GRVC Robotocs Lab. Universidad de Sevilla, ETS Ingeniería, Camino Descubrimientos, 41012 Sevilla (Spain)
asuarezfn@us.es

Valantis Tsiakos
Institute of Communication and Computer Systems, National Technical University of Athens 157 73 Athens
valantis.tsiakos@iccs.gr

Georgios Tsimiklis
Institute of Communication and Computer Systems, National Technical University of Athens 157 73 Athens
georgios.tsimiklis@iccs.gr

Patrick Ursolino
Quasset, Seinstraat 4, 1223 DA Hilversum
patrick.ursolino@quasset.com

Leonidas Valavanis
Inlecom Innovation, 145 61 Kifissia, Greece
valavanisleonidas@gmail.com

Dimitris Vamvatsikos
Resilience Guard GmbH, Switzerland
d.vamvatsikos@resilienceguard.ch

Goran Vasiljević
LARICS. Department of Control and Computer Engineering. University of Zagreb, Unska 3, Zagreb 10000 (Croatia)
Goran.Vasiljevic@fer.hr

Antidio Viguria
Advanced Center for Aerospace Technologies (CATEC), Wilbur and Orville Wright 19 – 41309 La Rinconada, Sevilla (Spain)
aviguria@catec.aero

Joaquín Borrego Villa
CATEC (Advanced Center for Aerospace Technologies) C/Wilbur y Orville Wright n19, 41309, La Rinconada (Seville), Spain
jborrego@catec.aero

Eleni Voulgari
Inlecom Innovation, 145 61 Kifissia, Greece
eleni.voulgari.inlecom@gmail.com

Jiaxu Xing
University of Zurich , Andreasstrasse 15, 8050 Zurich (Switzerland)
jixing@ifi.uzh.ch

Printed and bound by CPI Group (UK) Ltd, Croydon, CR0 4YY

15/09/2026

14962365-0002